Helielbia Alves Lucas
Márcia Santos Anjo Reis

# Health in science textbooks

Helielbia Alves Lucas
Márcia Santos Anjo Reis

# Health in science textbooks

ScienciaScripts

**Imprint**



Cover image: www.ingimage.com

This book is a translation from the original published under ISBN 978-3-330-76847-5.

Publisher:
Sciencia Scripts
is a trademark of
Dodo Books Indian Ocean Ltd. and OmniScriptum S.R.L publishing group

120 High Road, East Finchley, London, N2 9ED, United Kingdom
Str. Armeneasca 28/1, office 1, Chisinau MD-2012, Republic of Moldova, Europe
Printed at: see last page
**ISBN: 978-620-8-11369-8**

# SUMMARY

# INTRODUCED

The theme of this book is health in science textbooks. According to the National Council for Scientific and Technological Development (CNPq), the research falls within the area of human sciences, in the sub-area of health and biology.

I chose to work on the subject of health in the textbook because I was trying to bring together my current professional experience with my academic training in Pedagogy. Today, I work in the health sector, in a pharmacy, and although I don't have a direct and immediate connection with education, I saw an option to link the knowledge of my work with the degree course, taking advantage of the Course Conclusion Work (TCC) subject, offered in the last year, to deepen my knowledge of health.

It is essential to study the subject of health because of its great social importance. In this context, content related to healthy living habits, how to preserve the environment, how to prevent diseases, learning about hygiene, sexually transmitted diseases, drugs, among others, can be covered. In this way, they learn how to prevent and avoid contracting diseases, how to take care of their physical, mental and social well-being.

According to Sciliar (2007), until 1948, health was defined as the absence of disease. Over time and with socio-economic, cultural and lifestyle changes, the concept of health has changed. Since 1948, health has been defined by the World Health Organization (WHO) as "the state of complete physical, mental and social well-being and not merely the absence of disease" (SCILIAR, 2007, p. 1).

Analyzing the definitions of health presented above, we see that, when stating the meaning of health, mention is made of disease (illness), which leads us to realize that the concept of health is somehow interconnected with the concept of disease. Therefore, this book will use the expression health/disease, understanding health as the state of normality of the organism over a certain period of time, and disease as a disturbance in the functioning of the organism that can occur due to genetic, biological and external factors. Aware that no human being is totally healthy at all times of their life, in order to have and preserve a state of health, a number of factors need to be taken into account, such as good nutrition, decent living conditions, physical activity and the appropriate use of medication.

At the beginning of the 20th century, the subject of health was dealt with informally at school, in the subjects of hygiene, physical education and natural sciences, and the content covered explored how individuals can acquire illnesses; however, over the course of the century and realizing the importance of introducing the subject of health into school content, in 1971, with the Law of Guidelines and Bases of Education (LDB) No. 5.692, health content was formally included in the school curriculum, with the aim of encouraging children and adolescents to develop healthy habits in terms of "personal hygiene, nutrition, sports, work and leisure" (BRASIL, 1998, p.258), teaching how to preserve health as a whole.

In 1996, the National Curriculum Parameters (PCN) were created, and in this document some themes were raised that came to be called transversal, which due to the relevance and importance of the subject need to be discussed by different areas and not just by one discipline. Health was one of the cross-cutting themes proposed.

Assuming that the concept of health/disease has changed over time and that this topic should be addressed at school, at all levels and by all subjects, several questions can be raised: Does the school really address the subject of health? What concept of health/disease is explored during lessons? What teaching resources are used in the classroom to explore the topic of health? Do the textbooks for the initial grades of elementary school deal with issues related to health? Has the health/disease content included in the science textbook kept pace with changes in conceptions of health? Can a textbook adopt more than one conception of health/disease?

Faced with these questions, the main question of this book is: What is the conception of health/disease conveyed in the elementary school science textbooks adopted by the Jatai-GO municipal school system in 2014?

When it is said that the study will be carried out using textbooks, one might think that this is an outdated teaching resource that has already been widely exploited. It's worth noting that in some schools the textbook is still the only material used, which is why it continues to be the object of study.

According to Reis (2000, p. 11),

> Since the end of the 1970s, the textbook has been the subject of several studies, both national and international, which have resulted in academic papers, discussions and debates, whether university or not, and even journalistic articles that discuss its ideological and formal content with the general public. Various articles, books and research on these works have been published in recent years, and many criticisms have been leveled at them.[1]

In view of the criticisms levelled at textbooks in general - including the lack of information, the repetition of content which induces memorization, the compartmentalized content, activities with no link to the content, examples far removed from the students' reality, and the existence of conceptual errors (REIS, 2000), the MEC brought together a group of specialists to analyse textbooks from the 1980s onwards. The collections were analyzed and given one, two or three stars, depending on their quality. This policy has reduced criticism of the quality of textbooks. However, this doesn't mean that teachers don't need to analyze the collection to be adopted in their school, as they should check whether the content is in line with the reality experienced by the students, whether it is in line with the students' cognitive level, whether the suggestions for activities are consistent with the school's pedagogical proposal, among other things.

When studying the historical trajectory of the concept of health/disease, six existing conceptions were identified - the magical religious vision, the holistic model, the empirical-rational (Hippocratic) model, the Western scientific medicine (biomedical) model, the systematic model and the natural historical model of disease (processual model) (CRUZ, 2011). The hypothesis of this book is that the Science textbook colleague, adopted in 2014, is using the conception of health/disease that follows the processual model (natural historical model of diseases). In the case of a colleague published in the 2000s, it is hoped that the author of the colleague is keeping up with the most recent pedagogical reflections and proposals on the subject of health.

This book sets out to research the concept of health and to analyze how the subject is dealt with in science textbooks from the 2nd to the 5th grades of elementary

[1] Among these studies, Reis (2000) cites the following as theoretical references: Eco (1980), Pretto (1985), Freitag *et. All.* (1993), Fracalanza (1992), Faria (1986), Cicillini (1991).

school. The book for the first year of the class will not be analyzed due to the fact that in this grade, in the municipal network of Jatai, there is no science textbook in the classroom. However, it's worth noting that health content is introduced by teachers from early childhood education and also in the first year of elementary school, covering topics such as personal hygiene habits, eating habits, among others.

In order to achieve this general objective, specific objectives were defined that would contribute to the development of the research, including: finding out which collections of science textbooks have been adopted by the Jatai municipal school system for the initial grades of elementary school, identifying more than one didactic colleague, defining the criteria for selecting only one colleague from the science textbooks used by the municipal school system to be analyzed, investigating and theoretically researching the subject of health in order to discriminate between existing conceptions of health and their characteristics.

The research can be classified according to different criteria: the nature of its data, according to its objectives, in relation to the data collection procedures and, finally, according to the sources of information. Classifying it according to the nature of the approach, this is qualitative research, in which we sought to deepen our knowledge of the subject of health. Taking into account the general objective, the research is classified as exploratory, seeking to identify the concept of health adopted. According to the data collection procedures, the research is defined as bibliographical and documentary. Data was collected from books, theses and material available on the internet. The documentary research was carried out in the Science teaching collection entitled "Porta Aberta" (Open Door), by the authors Angela Gil and Sueli Fanizzi, whose object of investigation was the theme of health/disease. We used Cruz (2011), Balestrim and Barros (2009), the PCN for health (BRASIL, 1998) and others as theoretical references for the health theme. The authors Morh (1994) and Ilha *et all* (2013) were used to analyze the textbook. Ten criteria were defined for analyzing the collection used in the municipal schools of Jatai-GO in 2014, namely: scientific correctness; economic and geographical realities; adequacy to the minimum age; environment; depth; proposed activities; forms of proposed activities; eating habits; active lifestyles and reduction of health risks, which will be detailed in Chapter 4.

The book is divided into five parts: introduction, three chapters and final considerations. The introduction presents the importance of the topic, the research objectives (general and specific), the justification, the methodology and the stages of the book.

In the chapter "Health in Question", the definitions of health and illness are presented, as well as the history of health concepts and how the topic of health/illness is approached in education.

The chapter "The textbook in debate" highlights the importance of the textbook as a teaching resource, the state's relationship with the textbook, the creation of the National Textbook Program (PNLD), some criticisms of textbooks and the analysis criteria defined for science textbooks.

The last chapter, "Health in the Open Door Collection", describes the methodological path, the criteria used to analyze the Open Door Collection and then presents the results of the analysis of the health/disease theme in the textbooks for the 2nd to 5th year of elementary school, adopted in the municipal network of Jatai in 2014. Finally, the final considerations are presented.

# 1 HEALTH IN QUESTION

The purpose of this chapter is to show the changes in the concept of health that have occurred throughout history, to present the concept of health/disease in the 20th century and to explain how the subject of health/disease is worked on in education.

Health has its natural process altered due to a series of influences, causing suffering or public health problems. Sabroza (2004, p. 4) reports that "health/disease processes can be recognized, from the position of the observer and appears according to each position, as cellular alteration, suffering or public health problem", which is closely linked to the standard of living of societies, according to socio-spatial formation. The factors of production and consumption and social control are issues that can define many types of diseases, which generally occur in certain places and regions, accelerating the process of spreading diseases among people.

> The levels of expression of health-disease processes [...], at the individual level, can simultaneously be physiopathological alterations for the organic dimension; for the citizen, a representation and a role mediated by cultural values, and for the singular individual, suffering.
>
> At the level of society or complex socio-spatial formations like ours, they are expressed as public health problems, at the interface between state and society, between the private and the public, between the individual and the collective. And they can be analyzed as processes related to the reproduction of populations, as an element of the production and consumption processes, as crises and possibilities for social control, or as an issue related to human rights and citizenship (SABROZA, 2004, p. 5).

In our society, the state is responsible for providing health care to the population, building hospitals, health centers and distributing medicines; it is also responsible for funding research to discover methods of preventing and treating diseases. This is a right of the people that is provided for in the Federal Constitution of 1988, but which is restricted to paper, because we often read and hear reports in the media showing the inefficiency of Brazilian public health and the precarious conditions in which people are cared for in hospitals and health centers. We believe that if there is an improvement in society's living conditions (with access to basic sanitation, education and quality food), there will be a reduction in illnesses.

The transmission of knowledge on the subject of health/disease has been passed down throughout human history, from generation to generation, with the aim of helping to improve health conditions and broaden people's outlook on life. This knowledge includes basic conditions such as hygiene, healthy eating and basic sanitation. This knowledge needs to be included and incorporated into the family routine, right from childhood. As the National Health Curriculum Parameters (PCN) state,

> The health/disease process is inherent to life. Knowledge, pain and perplexities associated with illness, as well as recommendations for achieving longevity and physical and mental vigor, have been passed down from generation to generation throughout human history. Interpretations of the circumstances in which people protect themselves from illness, its causes, and accounts of its repercussions on the history of each individual and/or social group have always been present in different cultural formations (BRASIL, 1998, p. 249).

Through the advancement of science and technology, research is carried out and the results found contribute to improving the quality of life and health of individuals and society. These results include research into the prevention and treatment of diseases, drug formulations, vaccines, devices and surgical methods, among others. These discoveries are transmitted in the media and passed on by society to its generations. This does not mean that everyone has access to the information and products of science and technology, because, as the PCN for Health (BRASIL 1998) states, there is a lack of prioritization of social public policies.

> Humanity already has knowledge and technologies that can significantly improve people's quality of life. However, as well as not being applied to the benefit of all due to a lack of prioritization of social policies, there are a number of illnesses related to the genetic potential of individuals, or ethnicities, or to the sheer risk of living. No matter how good the living conditions are, people necessarily live with illnesses and disabilities, health problems and death (BRASIL, 1998, p. 251).

Nowadays, technology is important in investigating the causes of diseases, in methods that help prevent them, as well as in defining the treatment and cure of some diseases. Due to a number of factors, such as genetics, cultural and social issues, human beings are sometimes healthy and sometimes sick. According to the PCN on Health,

> The search for an understanding of the health/disease process and its multiple determinants leads to the conclusion that no human being (or population) can be considered totally healthy or totally ill: throughout their existence, they experience health/disease conditions according to their potential, their living conditions and their

> interaction with them (BRASIL, 1998, p. 251).

With this in mind, we understand that health must be considered as a whole. There are many factors that determine an individual's state of health/disease, including: their lifestyle (having good habits - eating habits, sleeping habits, exercising habits, spending part of their time on leisure, personal hygiene); not having any addictions; having good genetics, without many cases of hereditary diseases, and living in a place with minimal basic sanitation (treated water, garbage collection and sewage systems). These are some of the conditions for a healthy life, seeking physical, mental and social well-being. The PCN for Health highlights other factors that determine health/disease conditions

> Intricate mechanisms determine people's living conditions and the way they are born, live and die, as well as their health/disease experiences. The many factors that determine health conditions include biological factors (gender, age, personal characteristics that may be determined by genetic inheritance) and the physical environment (which includes geographical conditions, human occupation characteristics, sources of drinking water, availability and quality of food, housing conditions), as well as the socio-economic and cultural environments, which express levels of occupation and income, access to services aimed at promoting and recovering health and the quality of the care they provide (BRASIL, 1998, p. 251). 251).

In addition to the above-mentioned factors that condition the individual's health/disease, the care provided to the community must also be taken into account. Most of the time, when we talk about health, the first idea we have is how health care is organized, in other words, medical and hospital care and the condition of health posts. As Sciliar (2007) states,

> Medical care, outpatient and inpatient services and medicines are the first things many people think of when they talk about health. However, this is only one component of the health field and not necessarily the most important one; sometimes it is more beneficial for health to have clean water and healthy food than to have medicines (p. 1).

Without detracting from the great importance of medical resources, outpatient care and medicines, it is worth noting that diseases are often caused by a lack of hygiene habits and better living conditions among the population, leading to disease epidemics. With simple habits, such as taking care of garbage, animals and washing your hands, you can avoid various types of illness.

Society and the government, working in partnership, with the aim of improving personal and collective quality of life and concerned with preventing disease and not just curing it, can transform reality. The government has donated textbooks to schools with information on the subject of health/disease, which, when used correctly, can contribute to students' education. The topic of health is one of the subjects that should be dealt with in schools. This is not a recent concern. At first, the subject was covered by specific disciplines, including Hygiene, Physical Education, Biology or Natural Sciences, and was restricted to biological aspects, within the reductionist approach. According to the PCN for Health,

> Since the last century, although no specific space had been set aside to address the issue, health-related content has been incorporated into the Brazilian school curriculum in a way that reflected the same vicissitudes and perspectives with which these issues were dealt with socially. Thus, for example, subjects such as Hygiene, Dietetics, Physical Education and, more recently, Natural Sciences and Biology, disseminated knowledge relating to the mechanisms by which individuals fall ill or ensure their health. In its pedagogical practices, schools have systematically adopted a reductionist view of health, emphasizing its biological aspects. Even when considering the importance of the environmental conditions most conducive to the onset of illness, the relationship between the "sick person" and the "causal agent" continued - and continues to this day - to be prioritized (BRASIL, 1998, p. 257).

Over time, this reductionist view of health has changed. In 1997, the PCN defined health as one of the cross-cutting themes to be worked on in schools. With this proposal, all teachers, regardless of subject, need to address the topic of health, not as curricular content, but by working with the issues that students have doubts about, exploring the subject in different contexts and according to the focus of each subject, be it Biology, History, Physical Education, Geography, among others.

Considering this proposal to work on health as a cross-cutting theme and the government's policy of distributing textbooks to public schools, this research is justified, as it aims to analyze how health content is being approached in textbooks. Above all, because textbooks are often the only source of teaching resources for teachers in the classroom.

Understanding that over the years there have been transformations in the existing conceptions of health/disease and that they are constructed according to human needs, below we briefly present the models of health conceptions throughout history, as defined by Balestrim and Barros (2009) and Cruz (2011).

## 1.1 HEALTH/DISEASE CONCEPT

The concept of health/disease has changed over time. According to Cruz (2011), the conceptions of health/disease are: the magical religious vision, the holistic model, the empirical-rational model (Hippocratic), the model of Western scientific medicine (biomedical), the systematic model and the natural historical model of disease (processual model). Balestrim and Barros (2009) also provide a historical overview of the evolution of these conceptions of health/disease, describing in more detail how and why the changes that have taken place over the course of world political history have come about.

According to Cruz (2011), the oldest conception of health/disease is the **magical-religious view** prevalent in ancient times. In this conception, "relations with the natural world were based on a cosmology that involved gods and good and evil spirits, and religion in this case was the starting point for understanding the world and how to organize care" (CRUZ, 2011, p. 23). It was understood that illness was a punishment from the gods, and religion and belief prevailed. There were many transgressions due to lack of knowledge. There were no remedies or specific treatments for illnesses. "People conceived of the causes of illness as deriving from both natural elements and supernatural spirits" (CRUZ, 2011, p. 23). According to Balestrim and Barros (2009), this conception "attributes illness as the result of the action of external forces that invade the organism, causing illness due to some sin or curse" (p. 19). This conception is based on religious biblical studies which believe and assure that God can cure diseases and the devil is the cause of them.

Although this conception of health/disease is the oldest of all time, it is still considered by some to be true to this day. Pastors and religious leaders use biblical teachings to perform healings and miracles, according to people's faith.

Another conception of health/disease is the **holistic model**, whose notion of balance is responsible for its emergence in the 5th century BC. The cause of the imbalance (illness) was related to the physical environment, such as the stars, climate and insects.

> Health was understood as a balance between the elements and humors that make up the human organism. An imbalance in these elements led to illness. [...] According to this view, care should comprise the adjustment necessary to achieve balance between the body and the environment. The body as a whole. Ultimately, care means the search for health which, in this case, is related to the search for balance between the body and its internal and external elements (CRUZ, 2011, p. 23).

This idea of seeking total balance between the elements is not entirely wrong, as climatic issues, the physical environment and insects can have a major influence on the cause of diseases, as has been proven today. Studies continue to be carried out on this concept and other types of disease-causing agents have been discovered. This concept is still accepted today.

The **empirical-rational (Hippocratic) model** dates back to the 6th century BC. According to Cruz (2011), "Health in the Hippocratic conception is the result of the humors; illness is the result of their imbalance, and care depends on an understanding of these imbalances in order to achieve balance" (p. 24).

In a more advanced study in the quest to discover the causes of disease in Ancient Greece (460-370 BC), the physician Hippocrates presented his theory, the empirical-rational (Hippocratic) model, which debunked the idea that external factors were the cause of disease, thus identifying that the environment was directly related to the process, being responsible for causing certain types of disease.

> Hippocrates saw man as an organized unit and understood illness as the disorganization of this state. His work is characterized by the value of empirical observation, which was not limited to the patient himself, but to his environment (BALESTRIM; BARROS, 2009, p. 19).

The Greek Hippocrates' idea of medicine, according to Balestrim and Barros (2009), managed to influence Roman medicine and Claudius Galenus (129-199 AD) was one of his followers, being considered the main name in Roman medicine. Galenus believed that illness was caused by factors inside man himself (endogenous), originating from his physical constitution or lifestyle habits, which induced an imbalance, causing illness (BALESTRIM; BARROS, 2009). Still according to the authors, because the Roman Church did not allow new research, there was a period of regression in the advance of medicine at the beginning of the Middle Ages, returning to the idea that illness was caused by magical religious factors. However, due to the large number of epidemics that occurred at the end of the Middle Ages, "the causality of diseases was perceived as being caused by factors external to the individual's body" (BALESTRIM; BARROS, 2009, p. 20) and so research began again. The **Western or biomedical model of medicine** began in the 16th century and focused on the idea "that nothing should be accepted as true that cannot be identified as true"; in this context, medicine emerged (CRUZ, 2011, p. 24). The microbial theory proposed that each disease would have its own causal agent, so "the cause of disease was now a factor external to the organism, and man was the recipient of the disease" (CRUZ, 2011, p. 25).

Social revolutions took place against feudalism and a new mode of production emerged, known as capitalism. In this context, social, political and economic transformations took place and universities were created. They were important for increasing research, including medical research, which was aimed at unraveling the workings of the human body and finding causes and cures for illnesses. As Balestrim and Barros (2009) state,

> The end of the feudal mode of production in Western Europe was marked by a series of social revolutions directed against the traditional authorities of political and economic life, which ended up determining a new mode of production and, consequently, a new way of living in society: capitalism. The social, political and economic transformations, and the emergence of universities, led to a renewal of medical research, which until then had been the work of the clerics. (...) Moved by a new desire to discover the workings of the human body, doctors sought to explain illnesses through scientific studies and laboratory tests (p. 20).

In this context, it can be said that universities, through scientific studies and laboratory tests (analytical method), are partly responsible for the emergence of the Western scientific (biomedical) model of medicine. According to Balestrim and Barros (2009), the universities set about renewing medical research, which until then had been a function carried out by religious clerics. The analytical method began in the 16th and 17th centuries, a period known as the age of scientific revolution, and was fundamental to the development of scientific theories, despite the criticisms leveled at it: the fragmentation of thought, reductionism in science and the belief that complex phenomena can be understood when restricted to their constituent parts.

> With the advent of the modern age, traditionalism came under strong attack from rationalism; this began in the 16th and 17th centuries, in a period that became known as the "Scientific Revolution" [...]. Descartes was the founder of modern rationalism. His analytical method is probably his greatest contribution to science. It became an essential feature of modern scientific thinking and proved to be extremely useful in the development of scientific theories and the realization of complex technological projects (BALESTRIM; BARROS, 2009, p. 22).

For the authors, Descartes, responsible for founding modern rationalism through his analytical method, sought to understand phenomena by reducing them to their constituent parts until he arrived at a concrete answer, in other words, he led his thinking from the simple to the more complex.

With the scientific revolution, new technologies were developed, and with the creation of the microscope in the 17th century, etiological factors could be identified, and consequently diseases could be cured and even prevented, as stated by Balestrim and Barros (2009, p. 23):

> The invention of the microscope in the 17th century opened up areas previously inaccessible to the human body to observation, revealed the existence of disease-causing micro-organisms and made it possible to introduce serums and vaccines. It was a revolution because, for the first time, previously unknown etiological factors were being identified; diseases could now be prevented and cured, which meant a new way of producing truths, based on the logical operation of reasoning.

With the help of microscopes, it was proven that micro-organisms caused diseases, and through research, vaccines were invented to help prevent diseases and fight those that caused them.

Balestrim and Barros (2009) report that "in the 18th century, with the French Revolution and the consolidation of the industrial system, the concept of the social cause of illness appeared for the first time" (p. 24). This concept is in line with the **systematic model** described by Cruz (2011), who understands that the cause of illness is no longer just natural, but also social. Issues such as housing, work and hygiene conditions, which used to affect society's lifestyle, are now considered social causes that influence health and, consequently, are responsible for the manifestation of some diseases.

In 1976, the **model of the natural history of diseases** (HND) was systematized, defined by Cruz (2011) as the grouping of interactive processes that give rise to the pathological stimulus in the environment or another undetermined place, from man's response to the stimulus to the changes that lead to disability, invalidity, recovery or death. According to the author,

> The HND model aims to monitor the health-disease process in its regularity, understanding the interrelationship between the agent causing the disease, the host of the disease and the environment and the process of disease development. This form of systematization helps to understand the different methods of disease prevention and control (CRUZ, 2011, p. 27).

According to this conception, the development of the health/disease process of the entire cycle is presented in two sequential moments: the pre-pathogenic and the pathogenic. The pre-pathogenic period is the epidemiological period, which "concerns the interaction between the factors of the agent, the host and the environment" and the pathogenic period "is when man interacts with an external stimulus, shows signs and symptoms and undergoes treatment" (CRUZ, 2011, p. 27). For the author, the health/disease process presented in this concept allows us to consider the possibility of avoiding death, and this HND model brings different possibilities for preventing and promoting health.

It is worth emphasizing that the emergence of one model of health conception does not mean that the other has been abolished. Depending on the needs of society and its knowledge, different conceptions of health/disease can coexist at the same time. This is why I was interested in analyzing the concept of health present in the science textbooks adopted by the municipal school system in Jatai-GO for the initial grades of elementary school.

## 1.2 HEALTH/DISEASE THROUGHOUT THE 20TH CENTURY

Throughout the 20th century, there have been some changes in the concept of health/disease, in which each society has its own discourse on the body, worldview and social reality, and, according to these conditions, the population's way of living is determined. How society is organized can determine certain types of illness. According to Freitas and Martins (2008),

> Each society has a discourse on health/disease and the body, which is usually related to its worldview and social reality. We can see that there is a relationship between the way we live, get sick and die and social conditions, which allows us to relate living conditions, habits and illnesses (p. 2).

One concept of health that Freitas and Martins (2008) highlight in their article is that proposed by the WHO in 1948, which defines health as a state of physical, mental and social well-being, not restricted to the idea of the absence of disease, as previously determined. According to the authors, this is an advance on the Biomedical model (16th-17th centuries), which considered the human body as a machine. Different factors are now considered fundamental to achieving and maintaining health. In this context, health ceased to be an individual obligation and became a collective problem.

> The World Health Organization (WHO) presented a concept that contributed to progress in relation to the Biomedical (mechanistic/biological) model, which considered the human body to be a machine and health to be the proper functioning of the machine. According to this broader notion, "Health is the state of complete physical, mental and social well-being and not just the absence of disease". This new definition can be considered an advance in that it incorporates the mental and social as important to the individual. In this way, health becomes a collective responsibility and not an individual one, as it was seen from the perspective of the mechanistic/biological understanding, and many factors are now considered important for achieving and maintaining health (FREITAS; MARTINS, 2008, p. 2).

According to the authors, there has been an advance in the concept of health/disease, but this change was considered by them to be not very operative, "insofar as well-being is a subjective and non-measurable concept" (FREITAS; MARTINS, 2008, p. 2). In addition to the physical, mental and social well-being advocated in 1948, other factors came to be considered, including socio-economic, political, cultural and environmental dimensions, broadening and changing the concept and focus of the health/disease process. These dimensions were proposed in 1978 at the First International Conference on Primary Health Care, held in Alma-Ata.

> An important historical milestone in the development of a broader approach was the First International Conference on Primary Health Care, convened in 1978 by the WHO in collaboration with the United Nations Children's Fund (UNICEF), which took place in Alma-Ata, and which began to consider the socio-economic, political, cultural and environmental dimensions as fundamental to the maintenance and/or recovery of health (FREITAS; MARTINS, 2008, p. 3).

In 1986, at the Ottawa International Conference, one factor that was highlighted was the health discourse, which referred to issues such as health, food, shelter, education and a stable ecosystem, among others, which are considered determinants of quality of life and take into account socio-economic, political, cultural and environmental dimensions. According to Backes *et. all.* (2009),

> The idea of health as a quality of life conditioned by various factors, such as peace, shelter, food, income, education, economic resources, a stable ecosystem, sustainable resources, equity and social justice, emerged with the Ottawa International Conference in 1986 (p. 112).

Another important factor in the health/disease condition debated and defended at the Ottawa Conference is that the health of the individual and the community goes beyond the obligations of the health sector; it requires a healthy lifestyle and that they learn to take care of their health in order to improve it. They need to be able to identify their problems and meet their needs, adapting to the environment in which they live.

> The Ottawa Charter considers health to be a positive concept, for which personal and social resources and physical capacity are necessary. Thus, in order to be healthy, the responsibility goes beyond the health sector, as it requires a healthy lifestyle to achieve well-being. From this perspective, communities and individuals need to learn how to take care of their health in order to improve it. This should require individuals and groups to be able to identify their problems, satisfy their needs, change or adapt to the environment and, consequently, achieve well-being (BACKES *et. all,* 2009, p. 112).

In order to understand the health/disease process, it is necessary to follow the conceptual and theoretical changes related to the subject, analyzing and considering all the factors that contribute to the population's quality of life.

In 1997, the PCN was created and, in this document, a number of themes were raised that have come to be called transversal, which, due to the relevance and importance of the subject, need to be discussed by different areas and not just by one discipline. Health was one of the cross-cutting themes proposed.

> The inclusion of social issues in the school curriculum is not a new concern. These themes have already been discussed and incorporated into areas linked to the social sciences and natural sciences, with some proposals even building new areas such as the Environment and **Health**. The National Curriculum Parameters incorporate this trend and include it in the curriculum in such a way as to form an articulated whole that is open to new themes, seeking a didactic treatment that takes into account their complexity and dynamics, giving them the same importance as conventional areas. The curriculum gains in flexibility and openness, since themes can be prioritized and contextualized according to different local and regional realities and other themes can be included (BRASIL, 1997, p. 25, emphasis added).

Assuming that the concept of health/disease has changed over time and that this topic should be addressed at school, at all levels and by all subjects, several questions can be raised: Does the school really address the subject of health? What concept of health/disease is explored during lessons? What teaching resources are used in the classroom to explore the topic of health? Do the textbooks used in the early grades of elementary school deal with issues related to health? Has the health/disease content included in the science textbook kept pace with changes in conceptions of health? Can a textbook adopt more than one concept of health/disease?

This research focuses on the last three questions, which are related to textbooks and the subject of health. To do so, it is necessary to look at the relationship between education and the subject of health.

## 1.3 HEALTH AT SCHOOL

Concern about the issue of health can also be seen in the documents that define guidelines for Brazilian education. As the focus of this research is on the textbooks used in elementary school, from the second to the fifth year, in the Jatai municipal school system, this item looks at the presentation of the PCN, a document drawn up by the Ministry of Education (MEC)/Secretariat of Elementary Education (SEF) in 1997, with the aim of improving the quality of elementary school education and helping teachers to update their professional and teaching practices. It proposed working with cross-cutting themes: "Ethics, Cultural Plurality, Environment, Health, Sexual Orientation" (BRASIL, 1997, v. 9, p. 25). As can be seen, one of the cross-cutting themes proposed is Health, the subject of this book's investigation.

According to the PCN (1997), cross-cutting themes should focus on the construction and understanding of reality and the rights and responsibilities related to social and personal life. They should be worked on in existing areas or disciplines, relating them to important and urgent issues in society, and not as specific content for a given discipline. For example, the topic of Health should not be the responsibility of the Natural Sciences subject, but should be discussed when issues arise in the classroom, regardless of the subject being worked on. "The cross-curricular treatment of the theme is due precisely to the fact that it is approached in everyday school experience and not in the study of a 'subject'" (BRASIL, 1997, v. 9, p. 99).

The PCN on Health proposes general health objectives for primary education:

> - understand that health is everyone's right and an essential dimension of human growth and development;
> - understand that the condition of health is produced in relations with the physical, economic and socio-cultural environment, identifying risk factors for personal and collective health present in the environment in which they live;
> - to know and use individual and collective ways of intervening on factors that are unfavorable to health, acting responsibly in relation to health and the health of the community;
> - know how to access community resources and the possibilities of using services aimed at promoting, protecting and recovering health;
> - adopt self-care habits, respecting the possibilities and limits of their own bodies (BRASIL, 1997, v. 9, p. 101).

Health content, according to the PCN, should be "organized in such a way as to give meaning to its conceptual, procedural and attitudinal dimensions", helping towards a healthy life (BRASIL, 1997, v. 9, p.105). In order to select content for health teaching, certain criteria must be considered:

> - the importance of the process of growth and development in any particular living and health conditions for children and their social reality;
> - the most significant risk factors in the brazilian reality and in the age group of elementary school pupils;
> - the possibility of joint reflection on health promotion, protection and recovery measures;
> - the possibility of translating learning into personal and collective health care practices within the student's reach (BRASIL, 1997, v.9, p. 105).

Health content should provide students with a broader view of life, a more autonomous education, enabling them to establish their lifestyle, learn about their physical, psychological and social state, and develop habits for the individual and collective good. For Busquets *et all* (1999), the aim of teaching health/disease content consists of:

> To form autonomous personalities, capable of building their own lifestyles and achieving a balance that provides them with well-being, both physically and psychologically. To provide the means for children to become aware of their own physical and psychic states, their habits and attitudes towards the various situations of everyday life, and to build up an understanding of both the processes taking place in their bodies and the functioning of their personal and social relationships. To provide the means for children to get to know and use different ways of intervening in these organic processes, in order to change them for their own well-being (BUSQUETS *et all,* 1999, p. 65-66).

The authors define that children must know how their bodies work in order to acquire autonomy over their habits, learning about what is harmful and what is not harmful to their health, so that they can then choose the lifestyle they want to lead, knowing the risks that may occur. They should also learn about the changes that occur in the body as it grows and understand the care they need to take for their well-being.

The contents proposed in the PCN for Health (1997) are arranged in two blocks that indicate the individual and social dimensions of health, which are interrelated and have interconnections. The thematic blocks are: self-knowledge for self-care, and collective life. The block on self-knowledge for self-care proposes that students work on body hygiene habits, proper nutrition (learning about the nutrients each food contains, the care taken when eating certain foods and how they are prepared is important to avoid illnesses and, in some cases, malnutrition due to a lack of important nutrients for the body); the association between hygiene and nutrition needs to be emphasized. In the collective life block, the PCN for Health (1997) states that children should have the opportunity to identify problems in the interrelationships between health and the local environment and propose protective action, thinking of the community, learning about the diseases that are transmissible, how they are spread and the symptoms in order to know how to prevent them. It is important that there is a partnership with the health system and schools in informing people about vaccination campaigns so that more people are vaccinated. Work should also be done on accident prevention and, depending on the availability of the school, first aid skills should be taught, as well as information on the dangers of drug use and how to distinguish between different types of drugs and addictions. And, above all, teaching them how to live together, respecting differences of opinion and avoiding prejudice between those who are different, so that they are exercising their citizenship.

Some methodological proposals are put forward by the PCN for Health (1997), among them: the designation of health programs, with the aim of getting children and adolescents to develop healthy habits in terms of personal hygiene, nutrition, sports and leisure, seeking to teach methods for preserving personal (individual) and other people's (collective) health; seminars; campaigns disseminating information produced; promotion of debates; lectures with health professionals; visits to health service units; reading of information pamphlets made available at health service units; research on the internet and reading books.

The teaching guidelines presented by the PCN for Health suggest that: the content should be in line with the student's cognitive capacity, giving importance to the doubts that students have; there should be interdisciplinarity between the health/disease content and other subjects in the school curriculum; research should be carried out on the subject and the results published; there should be debate among students about events that are occurring and affecting the physical well-being of society, such as epidemics, tragedies, among others; and that the content should be in line with the students' social reality and that the teacher should make it appropriate to the situation experienced within the school.

Understanding that the knowledge about health/disease that students bring from home can differ from one another, depending on the social and cultural group, and depending on the point of view that each social group has about children, education and the health/disease dynamic, will determine the notion of health care. Maranhao (2010, p. 10) states that this health care "is based on knowledge, beliefs, values, skills built up and socialized in intergenerational relationships acquired through guidance and information provided by health professionals, educational institutes and the media". In this context, it is worth pointing out that

> Care as a professional practice is based on the exact, biological and human sciences

> and consists of attitudes and procedures based on up-to-date knowledge about the process of human growth and development and child health care (MARANHAO, 2010, p. 10).

The school must provide the conditions for teachers to be constantly informed and update their knowledge, so that they can learn to deal with situations related to childcare. This work should be carried out in partnership with other health professionals, including doctors, nurses, psychologists, occupational therapists, psychopedagogues and others, who can help with children's health care and the early diagnosis of possible illnesses. Maranhao (2010, p. 10) points out that "the teacher's close observation allows them to monitor children's development and reflect with parents both on their children's potential and on possible differences or difficulties that require special care".

Throughout this chapter, the importance of the subject of health inside and outside the classroom has been highlighted. Throughout history, different conceptions of health/disease have been constructed according to people's social and cultural needs. With the advance of science, technology and various research projects, new discoveries have been made, influencing the understanding, treatment and prevention of diseases in order to provide better health conditions. There is therefore a need to disseminate and socialize this knowledge, so that people not only learn what diseases are, but also the measures to prevent and treat them. Textbooks are one of these means of disseminating knowledge and discoveries, as well as providing information and helping to pass on knowledge. In this context, it is necessary that the information conveyed in textbooks is correct and in line with the student's cognitive level. For this reason, the next chapter will look at the importance of teachers working with suitable, high-quality material to help them pass on the health/disease content.

## 2 TEXTBOOK DEBATE

This chapter will discuss some of the reasons why textbooks are one of the most widely used resources in the classroom, the state/textbook relationship and the creation of the National Textbook Program (PNLD), the criticisms of textbooks and the analysis criteria defined for science textbooks to be analyzed and how the textbooks are classified.

The textbook is the main methodological resource used by teachers in municipal, state and private schools. There are other resources that can be used, such as video, magazines, the internet, newspapers, among others; however, due to a number of factors, the textbook continues to prevail as a source of research inside and outside the classroom. Mohr (1994) highlights the issue of lack of training and work overload to justify the prominence given to the textbook in the school space.

> [...] due to various factors, ranging from lack of preparation to lack of time to organize and plan the course, the teacher ends up adopting the textbook index as the syllabus for the school term. Throughout the school year, the textbook thus becomes the sole working tool for teachers and students: the textbook is the source of the information, texts and illustrations used in class; just as the exercises that are carried out in class or as homework come from the textbook (MOHR, 1994, p. 7).

In addition to these factors, others can be highlighted, such as low salaries, long working hours, lack of resources within the school and too many students in the same classroom. All these factors contribute to a lack of preparation and lesson planning, which consequently leads teachers to stick to the textbook, often adopting the summary presented in the book as their teaching plan.

According to Pimentel (2006), in order for textbooks to become useful material and be considered of quality for use in schools, they must offer adequate conditions to stimulate learning and contain indications that comply with Brazilian curriculum proposals. Furthermore, according to the author,

> The content should be accessible to the age group and cognitive development of the student. The text should encourage and value student participation during lessons, combating passive attitudes and behavior. The book should also promote integration between the various themes discussed in the chapters and value the experience and knowledge that the student brings to the classroom. The illustrations

> need to be up-to-date and correct and, whenever artistic resources involving colors, shapes and artificial dimensions are used, this should be clearly mentioned (PIMENTEL, 2006. p. 309).

Therefore, the content and activities proposed in the books must be in line with the students' cognitive level, without conceptual errors, encouraging active participation and valuing their prior knowledge; in addition, the figures, photos, graphs, charts and tables must be current, true and consistent with the reality of society, with information to facilitate community life.

In order to meet the demands of the job market, the state needs to educate society. To do this, it needs to guarantee the population access to education by increasing the number of places, building schools and, in general, qualifying the teaching staff. Over the years, there have been many changes in the field of education, and the state, as the funding agent for Brazilian education, initially took on the role of purchasing and distributing textbooks to public schools, which is still the case today. As well as buying the books, the government also started to evaluate their content. The relationship between the state and textbooks will be discussed below.

## 2.1 STATE AND TEXTBOOK RELATIONSHIP

According to Hoffling (2000), the program for distributing textbooks and materials by the Ministry of Education went through several stages, and was implemented by different agencies[2].

In 1938, the government created the National Textbook Commission, which established conditions for the publication and use of Brazilian textbooks. The Commission demanded the quality of the book, determining the control of information and correct language. It also created rules to restrict the import of textbooks in order to encourage Brazilian production.

[2] For more details on the evolution of the changes that took place in the bodies responsible for distributing textbooks in Brazil, read Hoffling (2000) - "Notas para discutao quanto a implementacao de programas de governo: em foco o Programa Nacional do Livro Didatico".

> The origins of the state/textbook relationship date back to 1938, when Decree-Law No. 1006 set up the National Textbook Commission, establishing conditions for the production, import and use of textbooks in Brazil. This decree established impediments to authorizing the publication of textbooks and requirements regarding the correctness of information and language (HOFFLING, 2000, p. 163).

In 1945, according to the author, by means of Decree-Law No. 8,460, the Commission's functions were resized and the State (at the federal level) became responsible for controlling the process of adopting textbooks in all the country's educational establishments. This function was decentralized with the creation of State Textbook Commissions in some states.

In 1967, the National School Material Foundation (Fename) was created, with the aim of "producing and distributing teaching materials to school institutions", without having the administrative organization and financial resources to carry out this task (HOFFLING, 2000).

From 1972 to 1975, the National Book Institute (INL) took responsibility, in partnership with publishers, for the coediting program of educational works, which became known as the Textbook Program (Plid), covering the different levels of education.

Under Decree 77.107/1976, according to Hoffling (2000), Fename underwent changes to its structure and became responsible for the activities of co-publishing programs for educational works, increasing the number of printed copies of textbooks and establishing a guaranteed market for publishers.

In 1983, the Student Assistance Foundation (FAE) was created, taking over the programs that were the responsibility of Fename, and in the same year, the Textbook Program (Plid) was incorporated into the FAE.

In 1984, the co-publishing system came to an end and the MEC became the buyer of the books produced by the publishers participating in Plid (HOFFLING, 2000, p. 163).

In 1985, through Decree-Law No. 91,542, the program's objectives were expanded and it was renamed the National Textbook Program (PNLD). The goal was "to serve all students from the first to the eighth grade in the country's federal, state, territorial, municipal and community public schools, with priority given to the basic components of Communication and Expression and Mathematics" (HOFFLING, 2000, p. 164).

As mentioned earlier, the FAE is a body linked to the MEC and is responsible for producing the Textbook Indication Manual, which is made available to all teachers so that they can choose the textbook that will be adopted by their school. The Manual contains a list of the books registered for the PNLD by grade and area, as well as an indication of the books and instructions on the criteria used in the selection.

To organize the Manual, the FAE publishes a notice and interested publishers submit their books for evaluation. The FAE then carries out the analysis, observing the non-disposable condition and physical aspects (paper quality, stapling, gluing, printing, cover, among others). The book's durability is taken into account, as each copy should last at least three years.

Once the titles have passed the non-disposable and physical aspects tests, they are included in the Manual. It is not the FAE's responsibility to correct errors arising from prejudice, stereotypes, printing, binding or stapling errors. If the FAE detects any conceptual inconsistencies or errors, it will return them to the publisher for corrections. When a conceptual error is found, it is more complicated to resolve because of its subjective nature.

After this evaluation, the Manual is put together and made available to schools so that teachers can choose the book they deem appropriate to adopt in their school over the next three years.

> The teacher is responsible for analyzing the titles presented, discussing them with his or her classmates and collectively recommending the books to be used in class. This must be expressed in a *form* (with a second choice of titles), sent to the Municipal Department of Education, which, via the State Department of Education, will forward it to the FAE for the purchase of the requested books (MOHR, 1994, p. 11, emphasis added).

According to the author, the school must fill in the form and send it to the FAE with two choices of books for the first to fourth grades and three choices for the fifth to eighth grades. Morh (1994) explains that the FAE's request to indicate more than one collection title per subject and from a different publisher is so that, if it is unable to negotiate the purchase of the first option requested, it will have the second option in hand to trade, with the aim of ensuring that students do not run out of books. In practice, this dynamic does not always correspond to what the Program recommends.

It is worth noting that the textbook distributed by the FAE does not belong to the student, it belongs to the school's "Book Bank".

> This means that, at the end of the school year, students hand in their books, which will be reused by new students in the following two years. Therefore, every year, the school has the right to request books for subjects and grades that have not been purchased in the last two years. This is specified in the *form* sent to each school. The FAE points out that failure to comply with this procedure renders the requesting school's application unfeasible (MORH, 1994, p. 10-11, emphasis added).

Various factors contribute to the PNLD being undermined, and Morh (1994) highlights some of them:

> These problems range from the lack of discussion between teachers in the same school to jointly select the textbook, to the chronic lack of funds from the MEC for its acquisition, which leads to significant delays in the distribution of textbooks (MORH, 1994, p. 11).

These problems reported by the author still occur today. The lack of good management, both in the purchase of textbooks and in their distribution, leads to a loss of teaching quality, since in many cases the distribution is so poor that the textbooks arrive at the end of the first semester of the school year. And when they do arrive, there aren't enough of them to meet the demand for a textbook for each subject required for each student, forcing them to stay in school cupboards and be used for two school terms. Another reason why children don't take their books home is the lack of family structure and the fear that the book won't be returned to school the next day, which would further damage teaching.

According to Mohr (1994), the FAE was abolished in 1996, and the execution of the PNLD was left to the National Education Development Fund (FNDE), which is a federal autarchy linked to the MEC.

It is worth noting that the MEC does not produce textbooks, but buys millions of textbooks from a small number of publishers. This situation leads Hoffling (2000, p. 164) to state that "the strong presence of private sectors - in this case, publishing groups - in the arena of decision-making and definition of public policy for textbooks can compromise the nature, the very concept of a social policy, with more democratizing contours".

Since 1996, the committee of experts hired by the MEC to evaluate textbooks has produced the "Textbook Guide", which is distributed to teachers in all schools, so that they can be informed and supported when choosing the textbook to be adopted in their school. The Textbook Guide states that,

> The textbook, like the Teacher's Manual, supports knowledge and teaching methods, and serves as a guide for the production and reproduction of knowledge. Therefore, it is essential that it stimulates further reading and presents a variety of bibliographical references, through different possibilities: specialized magazines, works available in libraries (school, city, higher education institutions, among others), as well as works and/or texts obtained through the World Wide Web *(Internet)*. Choosing a textbook is not a simple task, and it involves a great deal of responsibility. This guide will help you make a good choice. Look carefully at the characteristics that best meet your needs (PNLD, 2010 p. 8).

Even with other means, the textbook has prevailed, so it is necessary to carry out a critical analysis of the content of textbooks to find out if the information described is correct. Researchers are making a careful analysis of the content of textbooks and are reporting on their questionable quality.

## 2.2 TEXTBOOKS: CRITICISMS AND CRITERIA FOR BEING ANALYZED

The lack of efficiency in the content of textbooks means that teachers, increasingly insecure, are forced to produce their own teaching material, according to their pedagogical convictions. Exemplifying this situation, Megid Neto and Fracalanza (2003) comment on what primary school teachers are doing in relation to textbooks:

> Primary school teachers, on the other hand, have increasingly refused to faithfully

> adopt the textbooks on the market, as conceived and disseminated by authors and publishers. They constantly adapt the collections, trying to mold them to their school reality and their pedagogical convictions. They end up reconstructing the textbook they have adopted, which is not to their liking, given the effort put into such a reformulation without due professional recognition, nor is it to the liking of textbook publishers and authors, as they consider that these adaptations usually introduce errors and mistakes into the published works (MEGID NETO; FRACALANZA, 2003, p. 147).

According to the authors, teachers are working hard to meet the needs that the textbook cannot fulfill, and one of the problems is that the proposed content is not in line with the reality of school conditions. This situation does not please the teachers, who have to devote more time to preparing their lessons and do not receive the credit, nor the publishers and authors of the books, who claim that the adaptations made by the teachers often lead to conceptual errors.

According to Amaral and Megid Neto (1997), in 1994, the FAE commissioned an evaluation of textbooks from grades 1ª to 4ª and, in 1996, of textbooks from grades 5 to 8, which were carried out by a group of specialists from different institutes. According to the authors, the results have caused a stir,

> There has been a great deal of controversy between the government, publishers and education experts. More specifically, the press has reported that most of the textbooks widely used in Brazilian schools have been "disapproved" by the teams of evaluators. [...] this has occurred for a few main reasons: a large number of conceptual errors, the inclusion of prejudices of various kinds and inaccuracies in the illustrations; more rarely, deficiencies in the graphic aspect are pointed out (AMARAL; MEGID NETO, 1997, p. 13).

The authors claim that the textbooks contain conceptual errors, prejudices, incorrect illustrations and faulty graphics.

Textbooks are evaluated in various subjects, such as Portuguese, mathematics, natural sciences, etc. As the focus of this work is on the subject of health/disease, which is basically a subject covered in the subject of Science, the following is a report on the FORMAR-Ciencias group's research into the relationship between Science teachers and textbooks.

Megid Neto and Fracalanza (2003) report on a survey carried out by the FORMAR-Sciences Group at Unicamp in various cities in the state of São Paulo with around 180 elementary school science teachers to find out what role textbooks play in science teaching. The results of the survey were divided into three groups.

For the first group, "teachers indicate the simultaneous use of several teaching collections, from different publishers or authors, to draw up the annual planning of their lessons and to prepare them throughout the school term" (MEGID NETO; FRACALANZA, 2003, p. 148). The other group of teachers say that they use the textbook to support their teaching and learning activities. They point out that they use all the content in the textbook (texts, activities, illustrations, etc.) for in-class and out-of-class activities. And the third group admits that "the textbook is used as a bibliographic source, both to complement their own knowledge and for student learning, especially when carrying out so-called school bibliographic 'research'" (MEGID NETO; FRACALANZA, 2003, p. 148).

At the end of the survey, the teachers were asked to point out characteristics that they thought were important for a textbook to have:

> - Integration or articulation of the content and subjects covered;
> - Diverse texts, illustrations and activities that mention or deal with situations in the student's life context;
> - Up-to-date information and language suitable for the student;
> - I encourage reflection, questioning and criticality;
> - Illustrations with good graphic quality and visually appealing, compatible with our culture, containing captions and correct spatial proportions;
> - Experimental activities that are easy to carry out and with accessible material, without posing physical risks to the student;
> - Exemption from socio-cultural prejudices;
> - Maintaining a close relationship with official curriculum guidelines and proposals (MEGID NETO; FRACALANZA, 2003, p. 148-149).

According to the authors, the teachers described the same criteria required by the PNLD, which was published in 1994 in the guide for choosing textbooks and distributed by the FAE to primary schools.

The criteria are really important for the teacher to assess whether the book is good, and the government, aware of this need to make a good analysis, through the MEC, has released the "Textbook Guide", which covers not only primary education, but also secondary education and all subjects. In the area of Science, the book must be chosen according to two criteria:

> The first, considered *elimination* criteria for the collections, consist of incorrect concepts and basic information, methodological inaccuracy and inadequacy, damage to the construction of citizenship; the second set constitutes the *classification* criteria, which involve the adequacy of the content, proposed activities, integration between themes in the chapters, valuing the student's life experience, visual aspects of the illustrations and the teacher's manual (MEGID NETO; FRACALANZA, 2003, p. 149, emphasis added).

Even with the criteria defined specifically for the science area, the group of teachers surveyed by the FORMAR-Science team admitted that these criteria, because they are not totally specific, are also used when choosing other books, such as those for Portuguese and Geography.

According to Megid Neto and Fracalanza (2003, p. 150), "The guide published by the MEC in 1994 is the only one that is really concerned, albeit superficially, with the basic issues of science teaching, even though it was not considered in the assessment guides that followed it".

In 2010, the criteria that guided the analysis of science textbooks, described in the Guide to the Evaluation of Science Textbooks, were based on seven categories:

- pedagogical proposal;
- knowledge and concepts;
- research and experimentation;
- citizenship and ethics;
- illustrations, diagrams and figures;
- encouraging the use of other resources and media;
- Teacher's Manual (PNLD, Sciences, 2010, p. 10).

In the 2010 PNLD Call for Proposals, drawn up by the MEC, the specific features and characteristics of each category of analysis were described. The Guide explains how the books were evaluated and presents the evaluation form containing the results of the experts' analysis.

> In order to meet the demands of quality education, the textbooks enrolled in the PNLD are subjected to a pedagogical evaluation process based on **eliminatory criteria, common to all curricular subjects** and specific to each of them. These requirements cannot be breached if a work is to be purchased and distributed by the MEC. The common elimination criteria of the PNLD, 2013 are:
>
> I. respect for the legislation, guidelines and official standards relating to primary education;
> II. observance of the ethical principles necessary for the construction of citizenship

> and republican social coexistence;
> III.coherence and appropriateness of the theoretical-methodological approach taken by the work, with regard to the didactic-pedagogical proposal explained and the objectives pursued;
> IV.correcting and updating concepts, information and procedures;
> V.observance of the specific characteristics and purposes of the Teacher's Manual and adaptation of the Student's Book to the pedagogical proposal presented in it;
> VI.adequacy of the editorial structure and graphic design to the didactic-pedagogical objectives of the work (PNLD, 2013, p. 9, emphasis added).

In 2013, all the books analyzed were considered to be of good quality (PNLD, 2013). The justification found in the 2013 Science Guide is that, by meeting the qualifying requirements, the books can be considered suitable for classroom use.

These elimination criteria are considered for all subjects, but the most recent Science Guide (2013) includes them as analysis criteria:

> 1. proposals for activities that encourage scientific research through observation, experimentation, interpretation, analysis, discussion of results, synthesis, recording, communication and other procedures characteristic of science;
> 2. topics of study, activities, language and scientific terminology appropriate to the students' stage of cognitive development. [...]
> 3. initiation to the different areas of scientific knowledge, ensuring the approach of central aspects in physics, astronomy, chemistry, geology, ecology and biology (including zoology, botany, health, hygiene, physiology and the human body);
> 4. linking science content with other disciplinary fields;
> 5. he production of scientific knowledge as an activity that involves different people and institutions to which due credit must be given;
> 6. texts and activities that contribute to the debate on the repercussions, relationships and applications of scientific knowledge in society, seeking training for the full exercise of citizenship;
> 7. guidance for the development of feasible experimental activities, with reliable results and correct theoretical interpretation;
> 8. encouraging an attitude of respect for the environment, conservation and correct management;
> 9. clear and precise guidance on the risks involved in carrying out the proposed experiments and activities, in order to guarantee the physical integrity of students, teachers and other people involved in the educational process;
> 10. proposals for activities that encourage interaction and participation by the school community, families and the general public;
> 11. proposals for visits to spaces that favor the development of the teaching and learning process (museums, science centers, universities, research centers and others);
> 12. proposals for the use of information and communication technologies (PNLD, 2013, p 10).

The Science Guide (PNLD, 2013) makes it clear that even though science textbooks have improved significantly on some criteria, they need to improve on others. It has improved in terms of conceptual errors and content, but has not made progress in the use of technology, there is a lack of incentive to carry out experiments and there is no emphasis on presenting projects at science fairs. As the main focus of this book is health issues, it was decided not to delve into these criteria, which are covered in all the content, without detracting from the importance of each one.

After a careful analysis by the experts, a review is written with the opinion of each book or collection. Each Guide published from 1996 onwards presents a criterion for classifying textbooks, and the teacher is responsible for reading and interpreting these results and making their choice, according to the reality of their school and their students.

## 2.3 CLASSIFICATION OF TEXTBOOKS

In order to explain the criteria used to classify textbooks, we used Leao and Megid Neto (2006) as a theoretical reference, who explain that textbooks are analyzed according to the opinions of experts.

According to the authors of the 1994 document, "there is no classification of textbooks; there is a qualitative analysis and a classification table of the various indicators, filled in for each colleague, and it is up to the teacher to make the final judgment of the collections" (LEAO; MEGID NETO, 2006, p. 47). The books were analyzed and the opinions published, but it was up to the teacher to make the final choice of textbook.

From 1996 onwards, this opinion of the books analyzed was called the Guide. The 1996 Guide categorizes each book individually, classifying them as: Excluded, Not Recommended, Recommended with Reservations and Recommended. A single group of books could have different classifications, making it difficult for teachers who, by adopting a single group, were subject to having books of questionable quality to work with students.

- Excluded - books with conceptual errors, misleading, outdated, prejudiced or discriminatory of any kind;

- Not recommended - books in which the conceptual dimension is insufficient, with improprieties found that significantly compromise their didactic-pedagogical effectiveness;
- Recommended with reservations - books that had minimal qualities that justified their recommendation, although they presented problems that, if taken into account by the teacher, might not compromise their effectiveness;
- Recommended - books that satisfactorily met the common and specific analysis criteria used by the Program (BASSO, 2010, p. 5).

According to Leao and Megid Neto (2006), the Guides "adopt a prescriptive nature and classify books on up to three levels: Recommended, Recommended with Reservations and Recommended with Distinction" (p. 48). The recommended book receives three stars, the recommended with reservations two stars and the recommended with distinction one star.

When the Guide already offers an evaluation, "the problem is exacerbated by the fact that few collections receive the best evaluations in the Guides" (LEAO; MEGID NETO, 2006, p. 48). According to the authors, in 1998, only one collection received "three stars" and in 2000/2001, only two collections received such a rating. When a few collections are recommended (receive three stars), teachers tend to choose that colleague to adopt at school, and this ends up happening on a national level. This type of evaluation has created a national problem, as highly rated books do not serve as a good indicator for a particular region of the country. In agreement with Leao and Megid Neto (2006), textbooks should contain regional characteristics, and it is therefore up to teachers to be attentive when selecting textbooks, and it is not enough for them to have three stars.

Faced with the lack of good ratings for all the books in the same school, the 2002 Guide for 5th to 8th grade innovated by implementing the system of analyzing the collections. Leao and Megid Neto (2006) explain that teachers were adopting books from different collections in the same school, trying to select the best-rated ones, which led to discontinuity of content.

> The team responsible for the PNLD recognized the difficulty of evaluating individual books and, starting with the Didactic Evaluation Guide - 5th to 8th grades of 2002, began to evaluate the collection as a whole. Thus, this 2002 Guide, as well as the 2004 Guide -1ª to 4ª grades, already presents a single opinion for the

Science teaching collections (LEAO; MEGID NETO, 2006, p. 48).

In this sense, from 2002 onwards, the Guides began to evaluate and rate the colleague as a whole. The colleague is the one who receives the rating of Recommended, Recommended with Reservations or Recommended with Distinction.

Even with the colleague's evaluation, the problems continued, and from 2005 onwards the classification categories were abolished. The books that were evaluated were now "only labeled approved or recommended. That is, once approved, textbooks receive the same status" (BASSO, 2010, p. 6).

According to the author, from this year onwards the guides were no longer classified due to problems with publishers who were unable to sell their books to other companies in the private trade.

In 2010, the Science Textbook Guide included a table with the classification of the books. For each colleague's evaluation, two evaluators were responsible for analyzing the books, based on the criteria defined for science books. The collections had their covers removed and were given a code to guarantee their anonymity. In this way, the evaluators were able to judge impartially, avoiding giving preference to a particular author or publisher.

According to the PNLD (2010), the books were evaluated as follows:

> Each colleague was analyzed by two independent evaluators and, when necessary, by consultants from specific areas, using books without identifying the authors or publishers (uncharacterized books). The evaluators, listed at the beginning of this Guide, are active researchers in the fields of Science and Science Education. Once the individual analysis had been carried out, the two evaluators, who worked with the same colleague, met to complete their analysis. This was a moment of intense collective participation and exchange of conceptions and knowledge. A lot of dialog and important debates about the specificities of science teaching ensued in search of the best options. [...]
>
> Based on the analysis process adopted, the reviews of the approved collections and the general organization of this Guide were prepared (PNLD, 2010, p. 10).

At the end of the analysis, a general table is drawn up to give an overview of the classification of the collection by color. "The general table aims to give teachers a summary of the collections as a whole. The intensity of the purple color indicates the result of the evaluation of the collections: the more intense the purple color, the more the colleague meets the criteria specified in the notice" (PNLD, 2010,

p. 17). The reviews are as follows: the guide sets out questions relating to each criterion which the evaluators answer. When evaluating the book according to what the PNLD determines, the evaluators draw up a review detailing how each colleague is divided, and each chapter or unit is evaluated, giving a single opinion for the colleague.

To help teachers, the 2013 Textbook Guide includes a comparative table of the collections and, at the end, reviews of the collections as in the 2010 Guide, stating how they performed in each analysis criterion.

According to the PNLD for 2010, the colleague "Porta Aberta" (the object of analysis in this research) received an excellent evaluation, losing out only in terms of encouraging the use of other resources and media, as it did not encourage the use of computers, the internet and these new technologies. The same did not happen in the 2013 PNLD evaluation. In the overall picture, of the five criteria established, the Collection was predominantly purple only in two areas: pedagogical proposal and editorial project. In the other three areas - content; science and experimentation; and research and teacher's manual - the color was lighter, which means that it is not entirely in line with what is required.

Studies have been carried out and documents drawn up in order to improve the quality of textbooks, but issues such as teacher preparation are of the utmost importance if teachers are not to dismiss books classified as good quality because of traditionalism, lack of prospects and working conditions that are not in line with the reality of the school.

> [...] up-to-date works of good quality have already been published and, in practice, rejected by teachers, falling into oblivion or restricted use. We know that this happened because these were works that were not in tune with the programmatic and methodological traditions accepted by the teachers and, above all, because they were not suitable for their working conditions (AMARAL; MEGID NETO, 1997, p. 14).

We know that there are many problems in education that end up reflecting on the way teachers work. The textbook, as previously mentioned, is still one of the most used resources in schools, either because of the lack of financial conditions to acquire other types of resources or even because it is the most accessible and helps the teacher's work, presenting a sequence of pre-established contents and ready-made programmed activities, which make lesson planning easier, because they often work a double shift, or even because they have not had adequate training, and do not feel confident to draw up their own course plan. These factors don't justify the importance of the book, but they shouldn't be dismissed out of hand. As Amaral and Megid Neto (1997) report, due to a lack of working conditions and even academic training, teachers end up choosing books of questionable quality or not recommended by the Guides.

But it's worth pointing out that the textbook must meet minimum quality standards and the PNLD presents the analysis made by specialists in each area. Therefore, it is up to the teacher to follow the results published in the PNLD and select the colleague that best meets the reality of their school and students.

This is not an easy task, but it is a very important one, and we will be analyzing the health content of science textbooks in 2014. When we came across this study, we found confirmation that, although each school can work with the book of its choice, in the municipality of Jatal only one version is adopted for all schools.

# 3 HEALTH AT THE OPEN DOOR COLLEAGUE

This chapter will describe the methodological path of the research, present the criteria used to analyze the Open Door Collection and analyze the textbooks for the 2nd to 5th grades of elementary school, in an attempt to identify how health/disease is approached.

## 3.1 METHODOLOGICAL APPROACH

Having decided that the specific study would be to analyze how the health/disease content is being approached in the science textbooks of the initial grades of elementary school in the municipality of Jatai, the first step was to identify which science class was adopted in the schools in 2014, i.e. the document to be investigated.

To do this, it was necessary to go to the Municipal Department of Education to find out which collections of science textbooks were adopted in 2014 by the school system. If the adoption of more than one collection was identified, a survey would be carried out to find out which collection was the most accepted by the municipal schools and then the analysis would begin. Normally, every three years, schools choose the textbook that will be adopted for the next three years and can opt for different collections.

An appointment was made with the municipality's pedagogical coordinator at the Department of Education to request a list of the collections purchased for the municipality's schools in 2014. According to the municipal pedagogical coordinator, in 2014, due to problems with the distribution of books in previous years and due to the quantity of books to be purchased, the municipality opted to adopt a single collection for all schools, the Open Door Collection, by FTD. The authors of this collection are Angela Bernardes de Andrade Gil[3]and Sueli Fanizzi[4]and the 1st[a] edition was published in 2011.

[3] She has a degree in Languages and is a Portuguese Language and Science teacher at public and private elementary schools in Sao Paulo.

[4] She has a Master's degree in Education from USP, a degree in Pedagogy, and is a teacher of higher education in the early years of elementary school in the private school system in Sao Paulo.

Once the colleague has been identified, it is necessary to present the analysis criteria that will be adopted to carry out the study.

## 3.2 ANALYSIS CRITERIA

When looking for texts that provided theoretical support on the subject of health/disease, we found some research and/or texts that explored existing conceptions of health and analyzed the subject of health in textbooks. After reading and studying these texts, we decided to adopt the analysis criteria presented by Adriana Mohr (2000) and Phillip Ilha *et. all.* (2013), who also analyzed the theme of health in textbooks. After drafting the subtitle referring to the PNLD, written in the previous chapter, we found that some of the analysis criteria listed by Mohr (2000) and Phillip Ilha *et. all.* (2013) follow the criteria adopted by the PNLD for Science.

The criteria are important to guide the analysis. They make it possible to base the assessment that will be made on the books. In all, 10 criteria were defined for analysis in this research.

The first criterion observed was the **presence or absence of a concept or definition**. It was checked whether the concept or definition presented was clear and explicit enough to understand what was being explained. According to Mohr (2000, p. 90), "when talking about conceptualization in a textbook, it is understood that the text should present information and explanations developed in such a way as to allow the student to gain a general understanding (albeit imprecise and not formalized) of the subject in question".

The **economic and geographic realities** criterion looked at whether "the content and the way it is presented identify with situations and experiences lived by the student" (MORH, 2000, p. 92), adapting to the diversity of realities, both economic and geographic. During the analysis, it was assessed whether the content presented addresses issues that can be explored and adapted to the economic and geographical realities of Goias, where the municipality in which the textbook was adopted is located.

Even though the textbooks are delivered by the government and the content is intended to be used throughout the country, it must make it possible to explore regional diversities, and the teacher must be attentive when analyzing the books and making his or her choice to adopt a particular colleague.

Another criterion used by Mohr (2000) and which was also used in this research is whether the **content** presented in the selected collection of textbooks is **appropriate to the cognitive level of the students**, i.e. the minimum age. According to the author's analysis in her research, she observed flaws and certain inconsistencies in the textbooks with regard to the proposed content and the minimum age that students should have to learn certain subjects. In this sense, we decided to analyze whether the topic of health/disease is being dealt with at the student's cognitive level.

Morh (2000) says that the content of health is directly linked to external factors, one of which is the environment. Therefore, "it is of fundamental importance that health programs take an ecological approach, including man as part of the ecosystem" (p. 92). In this context, the **environmental factors** that influence health/disease were considered in this research, observing the role of man in transforming the environment.

A textbook colleague presents a series of contents and activities that are distributed among the volumes, providing a sequence, according to the grade that will be used. Understanding that the content should be in-depth and have continuity from year to year, we looked at whether the Porta Aberta colleague is concerned with **in-depth content**, which was considered as one of the analysis criteria.

Another criterion was the **activities** proposed in the textbook. Morh (2000) comments that "It is desirable that the proposed activities include the acquisition of knowledge, the ability to analyze, criticize and stimulate the students' initiative" (p. 92). The author identified the types of activities in five categories:

> [...] those that require the formation of a concept or understanding of the text; those that suggest solving problems or proposing solutions; those that correspond to literal questions (copying); those that presuppose activities outside the textbook and those in which the student must express an opinion or report on behavior and experiences. It also identified two ways of carrying out the proposed activities: individually or in groups (MORH, 2000, p. 92).

According to Mohr (2000), one of the determining factors in assessing the quality of textbooks is the form and type of activities proposed in them for students to solve. In this way, the types and forms of activities proposed in textbooks related to the health/disease theme were analyzed, observing whether they arouse the acquisition of knowledge, the ability to analyze, criticize, take a stand and take initiative on the part of the students.

The **eating habits** criterion was based on Ilha *et. all.* (2013), who state that texts or keywords related to food and/or nutrition should be based on improving the quality of life and promoting the health of individuals and society. Therefore, during the analysis, we tried to identify in the chapters, texts, activities and/or illustrations of the themes of food and nutrition how they are approached, observing what focus is given.

Another criterion established was **active lifestyle**. According to Ilha *et al.* (2013), texts or words related to physical activity should be health-oriented or address recommendations for practicing health-related physical activities. During the analysis of the books in this research, we looked at how the texts describe the issues related to practicing physical activities, observing whether they are aimed at improving quality of life and preserving health or just for leisure.

During the research, the **health risk** criterion was also taken into account, looking for texts that report on the topic, observing whether they clarify the risks and dangers that addictions cause for health conservation and the precautions for preventing addictions. In the research carried out by Ilha *et. all.* (2013, p. 110), one of the criteria used in the analysis was the reduction of health risks, looking for "texts or keywords related to attitudes that avoid illness, accidents, drug use, smoking and drinking", which were also considered in the analysis of this research.

The last criterion for analysis was the **concept of health** adopted in the books. It was analyzed whether the content covered on health/disease makes it clear what the author's conception of health is, whether it is based on more than one conception, what he defines about the importance of working with health/disease in the classroom.

The next section analyzes the books of the Science class selected as the object of investigation in this research, based on the criteria (categories) presented.

## 3.3 ANALYSIS OF THE OPEN DOOR COLLECTION

As previously mentioned, this study analyzed the science textbooks from the 2nd to the 5th grades of the Open Door Collection. It was decided to analyze the books separately, i.e. one book at a time.

Table 1 below shows the 10 criteria used to analyze the books in the Collection and examines what each book has, doesn't have or is undefined (vague, generic). The comments on these observations can be found in the analysis of each book in the following subtitles.

**Table 1** - Analysis criteria and what each book presents

| Criteria (categories of analysis) | 2nd Year | 3rd Year | 4th Year | 5th Year |
|---|---|---|---|---|
| 1. **Presence or absence of conceptualization or definition** | Features | Features | Features | Features |
| 2. **Economic and geographical realities** | Features | Features | Features | Features |
| 3. **Suitability for the minimum age** | Features | Features | Features | Features |
| 4. **Environment** | Features | Features | Features | Features |
| 5. **Deepening** | Undefined | Features | Features | Features |
| 6. **Proposed activities** | Undefined | Undefined | Features | Features |
| 7. **Forms of proposed activities** | Undefined | Undefined | Features | Features |
| 8. **Eating habits** | Features | Not shown | Features | Features |
| 9. **Active lifestyles** | Not shown | Not shown | Not shown | Features |

| 10. **Reducing health risks** | Features | Not shown | Not shown | Features |
|---|---|---|---|---|

Table 2 shows the number of units each book contains and shows how many of them specifically explore the topic of health and how many of them present health-related approaches.

**Table 2 - Number of units in each book, how many are health-specific and explore the subject.**

| | 2nd Year | 3rd Year | 4th Year | 5th Year |
|---|---|---|---|---|
| **Number of units of the book** | 9 | 9 | 9 | 9 |
| **Specific health units** | 1 | 1 | 0 | 1 |
| **Units dealing with health/disease** | 2 | 5 | 5 | 6 |

### 3.3.1 Analysis of the 2nd grade textbook

The summary of the 2nd grade book is divided into 9 units: Perceiving the world (Sense organs, Sight, Hearing, Touch, Smell, Taste); Preventing diseases (Diseases: Symptoms and prevention, Common diseases, Viruses, And long live vaccines, Vermin, Diseases caused by bacteria); Living and non-living beings (Identifying living and non-living beings, Living beings, Different environments); Animals of all kinds (How animals are born, Oviparous animals, Viviparous animals, Ovoviviparous animals, Metamorphosis, Feeding animals, The bodies of animals, Ways of getting around, Endangered animals); The plant world (Plant reproduction, The parts of a plant, Plants also eat, Plant environments); Water, air and soil (Where do we live, Water, What is air?, Taking care of the environment we live in (Brazilian environments, Causes of deforestation, Reforestation, Conservation areas, The environment of cities, Preserving the environment of my city); Transforming materials (Products of plant origin, Products of animal origin); Planet Earth and other stars (The Earth, our home, The Solar System, Size of the Earth, Where is our Solar System? Shine, shine star, Sun, our great star, Next stop: Moon!, Phases of the Moon, Day and

Night).

Of the nine units, only one title makes direct reference to the theme investigated in the research, entitled - Preventing illness. However, after analyzing the contents of the nine units, it was observed that two other units deal with the theme: Perceiving the world (unit 1) and Water, air and soil (unit 6).

In Unit 1 - Perceiving the World, they work on the sense organs (smell, taste, sight, hearing and touch). They explore the functions of each sense organ, the care that should be taken to protect them and avoid illnesses, the symptoms that can appear when the sense organs have a problem, guidance on speaking to the teacher when you identify a problem or seeking help from a doctor (specialist), respect for the disabled, the advances in science and technology to alleviate problems linked to the sense organs (hearing aids, glasses).

In unit 6 - water, air and soil, the health-related topic explored was polluted air and the incidence of respiratory diseases. No work was done on polluted or contaminated water, water-borne diseases, the importance of water treatment for human health, or how to avoid polluting the soil and soil-borne diseases. Unit 2 - Preventing diseases, explains the symptoms and ways of preventing some diseases that are considered common among children, discusses the invention of the microscope and its usefulness for observing tiny beings, and explains about viruses (symptoms and ways of spreading them). It also discusses the importance of vaccinations for the prevention of diseases and shows an illustration of a vaccination card of a child who has had all the vaccinations and asks everyone to compare their card. This is a way of putting into practice the theory studied in class. The images and text make the content clearer.

Also in this unit, there is a short clarifying text that explores worms and how to prevent them. In another topic, it describes the diseases caused by bacteria, and gives the example of caries, which is the result of the presence of bacteria, and takes the opportunity to give tips on good brushing, explaining with pictures how teeth are formed, how to brush well and use dental floss. The unit has the topic HEALTH TIP, and the tip is:

> It is possible to prevent diseases. In addition to vaccinations, eat fruit, vegetables, cereals, lean meat, fish and poultry, eggs and milk. Drink 8 to 10 glasses of water a day. Play sports, have good hygiene and don't forget to brush your teeth!
> Healthy eating creates defenses in the body to protect against viruses (GIL; FANIZZI, 2011, p. 42).

Paralleling the content proposed in the 2nd grade textbook with what is proposed in the PCN for Health, it can be seen that they are in agreement. The PCN for Health emphasizes that

> The acquisition of body hygiene habits begins in childhood and the importance of practicing them systematically is highlighted. Experiences of doing procedures together with the child that can be carried out in the school environment, such as washing hands or brushing teeth (BRASIL, 1997, p. 107).

By ensuring that students learn and reflect on hygiene habits (personal and food), they are ensuring the prevention of various diseases that are caused by viruses and bacteria, which can be avoided with hygiene, as was discussed on pages 40 and 41, when the authors teach tips for brushing and flossing.

Starting from the selected analysis criteria, we began by looking at the **presence or absence of conceptualization or definition.** In the 2nd grade book, the concepts presented are very clear and scientifically correct. To illustrate, here is the conceptualization of worms in the book:

> Vermin are diseases caused by worms. They enter our body when they are still small or in the form of eggs, through contaminated food or water, or when we put our dirty hand in our mouth (GIL; FANIZZI, 2011, p. 36).

Scientific terminology is used so that children become accustomed to and incorporate scientific expressions into their vocabulary, believing in the students' potential for understanding and comprehension.

When analyzing the **economic and geographic realities** criterion, it can be seen that in the 2nd grade book the authors pay attention to this issue. They present illustrations depicting the everyday situations of children from different regions of Brazil and ask the children to get together in groups and discuss the guidelines that should be given to prevent worms. This activity mentions different regions, cultures and economic classes, seeking to promote the reflection of different national realities in the classroom, not specifying just one region, but covering different regional characteristics, such as the sertao of Ceara, the state of Minas Gerais, an indigenous village in Mato Grosso, the interior of Rio Grande do Sul and Para. Although it doesn't mention Goias, some examples are present in the reality of the state and can be explored by teachers.

With regard to the criterion of **suitability** for the **minimum age,** it is understood that the content is suitable for children in 2nd grade. In the first unit - Perceiving the world, as previously mentioned, students are introduced to the five sense organs, making them realize the importance of each organ in their daily lives. Overall, the book provides a lot of information on the subject of health/disease in a way that is understandable to the age group of the students, but does not go into too much depth so as not to complicate the children's understanding. This unit provides tips for parents to observe their children and identify if they have any listening difficulties. The explanation is easy, appropriate to the cognitive level of the students and they can pass on this information to their parents or even take the test with their siblings.

> Parents can identify if there is a problem with their baby if they are not startled by the sound of a door slamming or the loud noise of an object falling to the floor. If you can't hear very well, talk to your teacher. He or she will be able to advise your parents or guardians to refer you to an ENT specialist (GIL; FANIZZI, 2011, p. 20).

When analyzing the **environmental factor** criterion, we see that the 2nd grade textbook talks about deforestation, the importance of water, teaches about soil, explores the issue of animals that are going extinct, the environment in general; it also addresses changes that man causes in nature due to his actions, and, above all, tries to teach children the means to reduce the consequences of the problems that man has caused to the environment.

With regard to the environmental factor and the theme of health/disease, unit 6 - Water, air and soil - could have explored various subjects linked to human action and the pollution of water, air and soil and the impact on individual and collective health in the regions affected. But it restricted itself to mentioning some respiratory diseases caused by polluted air.

The textbook for the second year was very superficial, as Mohr (1994, p. 20, emphasis added) states; the focus of the content should have dealt with issues related to preventive measures, and at the same time, "offer subsidies for understanding *curative* procedures", recommending actions that take "into account both individual and collective behaviors".

When analyzing the criterion of **depth of** content, as this is the first book of the colleague studied, it is not possible to make any statements at this point, without first having analyzed the other books of the colleague. For this reason, the expression indefinite has been included in Table 1. In the next books analyzed, this criterion can be commented on, as there was a parameter for comparison, i.e. it can be observed whether the theme has been deepened with each passing year, "whether the main themes recur in the other volumes of the colleague" (MOHR, 1994, p. 20).

In the **activities** and **forms of activities proposed** by the authors, the majority are for content fixation, which is why the word undefined is included in Table 1, as the criterion envisages activities that awaken the ability to analyze, criticize, take a stand and take initiative. Some group activities are also suggested, but to a lesser extent. Some units have a topic which suggests that students carry out experiments relating to the content and then explain and report on their experience.

In units 1, 2 and 6, which deal with the topic of health/disease, few questions lead to reflection on prevention; the majority are content fixation questions, i.e. the text comments on the subject and the activities require memorization. There are a few exceptions, such as in unit 1, in which there is a question that explores sponges that we should avoid so as not to irritate our eyes (p.11); and in unit 2, the authors ask a series of questions related to the individual health of each student, in which they have the opportunity to reflect on what they have studied and its applicability in their lives. They ask if the student has been to the health center to get vaccinated, so that he can pick up his vaccination card and check that his vaccinations are up to date.

In terms of **eating habits**, the unit "The plant world" presents the parts of plants, telling us how plants reproduce, whether through flowers or seeds. The topic does not delve into the importance of these foods in daily nutrition, and the authors only focus on the formation of plants. Since the PCN for Health (1997) suggests that food can be a way of preventing illnesses, "adequate nutrition is another essential factor in growth and development, in the performance of daily activities, and in the promotion and recovery of health" (p.107), it should therefore be more specified in this topic, because no opportunity should be missed to teach how to eat properly for good health.

With regard to the **active lifestyle** criterion, this volume didn't identify any units that encouraged sports practice. In unit 2 - Preventing illnesses, there is only one section of text, in the health tip, which mentions that you should practice sport, but it doesn't go into the issue in any depth, nor does it report at any point on the importance of physical activities for preventing illnesses and maintaining health and quality of life. The text contains the following health tip:

> It is possible to prevent diseases. In addition to vaccinations, eat fruit, vegetables, cereals, lean meat, fish and poultry, eggs and milk. Drink 8 to 10 glasses of water a day. **Play sports**, have good hygiene and don't forget to brush your teeth!
> Healthy eating creates defenses in the body to protect against viruses (GIL; FANIZZI, 2011, p. 42, emphasis added).

According to the PCN on Health (BRASIL, 1997), the factors that can cause risks to children's physical integrity are "communicable diseases, domestic or traffic accidents (being run over) and those resulting from social life - ill-treatment, accidents with firearms, sexual violence or health problems associated with child labor" (p. 113). Under the criterion of **reducing health risks**, in Unit 1 - Perceiving the World, a situation is described in which the sense of smell can help to perceive a fire situation by smelling the smoke and seeking help in time to avoid further problems (p.23).

The sixth unit - Water, air and soil - explains the importance of water for our bodies and how water should be treated so that it can be consumed (p. 95-96). Problems such as dehydration and air pollution are also mentioned, but the damage these problems can cause to health (both individual and collective) and to the environment is not discussed in depth. In this unit's health tip, it appears:

> Our body is made up of around 70% water. There is even water in our bones.
> Water reaches our bodies through the liquids we ingest. That's why it's important to

drink water every day, at least eight glasses a day (GIL; FANIZZI, 2011, p. 99).

When analyzing the criterion of reducing health risks in the 2nd grade book, it was found that the texts prioritized a descriptive approach to some diseases and short reports on their symptoms or how to prevent them. As stated by ILHA *et all*.(2013, p. 113),

> In order for students to promote their own health, it is not enough to know about diseases and their prophylaxis, but it is necessary to develop reflections that contribute to the construction of individual and collective actions that promote quality of life and, by extension, individual health and that of the community in which they live.

The authors criticize the superficial way in which books deal with the issue of health/disease and believe that the process should be worked through in a reflective way to raise awareness of the importance of care for a better quality of life.

### 3.3.2 Analysis of the 3rd grade book

The Open Door Collegiate Science book for 3rd grade has 9 units, including: Earth and the Universe (Studying planet Earth, Earth's movements, The seasons, Light and shadow, The solar system); Wind: a source of energy (Air in motion, Wind formation, Wind direction, Wind speed, Wind force); Importance of soil (Soil formation, How is soil studied? Soil permeability, Soil pollution); Water and health (Water goes... water comes, Water cycle, The birth of a river, Water and health); Separation of materials in a mixture (Materials that mix completely, Materials that don't mix completely, Water: Universal solvent, Separating mixtures); Preserving the environment (Basic sanitation, Water supply and treatment, Garbage collection, Garbage disposal, Citizen's rights and duties); Studying living beings (Classifying living beings, The kingdoms, Neither the plant kingdom nor the animal kingdom, Animal and plant products); How living things reproduce (The beginning of a new animal life, Human reproduction, Caring for babies and young, The beginning of a new plant life); Getting to know the human body (The human body, Parts of the human body, Human types, People with disabilities, Senior citizens).

Of these 9 units, only one specifies the theme of health in its title, unit 4 - Water and Health. However, when analyzing the contents of the other units, it was observed that five of them deal with issues related to health/disease, which are: unit 5 - Separating materials from a mixture; unit 6 - Preserving the environment; unit 7 - Studying living beings; unit 8 - How living beings reproduce; unit 9 - Getting to know the human body.

In Unit 4 - Water and Health, some issues related to the topic of health/disease, the subject of this research, are presented, such as: diseases that humans can contract when in contact with contaminated water (leptospirosis, cholera, schistosomiasis, dengue), explaining how they are transmitted, what the symptoms are and how to prevent them, and the advances in science, talking about the vaccine against cholera and diarrhea (p.65-67).

Looking at Unit 5 - Separating materials from a mixture, one might think that this subject has no connection with the health/disease theme, but it is important to note that this theme is not restricted to human beings, but to living beings. In this unit, the authors describe the relationship between the oil spill in the sea and living beings (animals and seaweed) (p.75). They also explain the processes of separating mixtures, drawing a parallel with the water treatment plant, highlighting evaporation. It cites the extraction of sea salt by the process of evaporating sea water as an example and gives health tips on the importance of iodized salt and the need to consume it in moderation (p. 87).

During the analysis of the content covered in unit 6 - Preserving the environment, it was found that subjects related to water and sewage treatment and their relationship to health, the diseases that affect people who work in garbage dumps, selective garbage collection, and the rights and duties of citizens in relation to basic sanitation (p.99).

In Unit 7 - Studying living beings, the authors comment on the types of fungi and the diseases they cause in humans and plants (p.112) in the subtitle Neither the plant kingdom nor the animal kingdom.

In the next unit, 8 - How living beings reproduce, the authors highlight how the fetus feeds in the intrauterine environment and the newborn in the first months of life, emphasizing its importance for health (p.124).

And finally, in Unit 9 - Getting to know the human body, the book addresses issues related to the right to education and work of people with disabilities, the difficulties and prejudices they suffer in everyday life, the importance of physical activity, citing the Paralympic Games and the physical and mental transformations that occur in human beings, highlighting old age (p.149).

For science teaching, all the content included in this book is very important for students' learning and knowledge, but our focus is on the topic of health/disease. Studying the PCN for Health (1997), we found that it suggests some content to be worked on in order to explore the theme:

> - knowledge of the resources available to children (activities and services) for the promotion, protection and recovery of their health, the possibilities of using them and the ways of accessing them;
> - ways of participating in collective actions accessible to children in their community;
> - knowledge of the vaccination schedule and their own vaccination status;
> - **the main signs of symptoms and the most common communicable diseases in the student's reality, forms of contagion, prevention and early treatment to protect personal health and that of others**;
> - problems caused by the use of drugs (smoking, alcohol, narcotics);
> - knowledge of basic safety rules when handling instruments, in traffic and when practicing physical activities;
> - simple first aid measures for: abrasions and contusions, convulsions, animal bites, burns, fainting, insect bites, sprains and fractures, drowning, poisoning, cramps, fever, electric shock, nosebleeds, diarrhea and vomiting, traffic accidents;
> - **the most significant environmental factors for health present in children's daily lives: water treatment systems, ways of disposing of human and animal waste, garbage** and agrotoxics;
> - mapping the necessary changes in the environment in which you live;
> - **links between environmental preservation and restoration and improved quality of life and health**;
> - rejection of acts of destruction of environmental balance and health;
> - active participation in maintaining a clean and healthy environment at home, at school and in public places in general;
> - **solidarity with the health problems and needs of others through attitudes of help and protection for disabled** and sick **people** (BRASIL, 1997, p. 114-115, personal emphasis).

Drawing a parallel between the content explored in the 3rd grade book and the content proposed by the PCN on Health, it can be seen that the authors have complied with four of the Parameter's suggestions, which are in bold in the previous quote.

Based on the analysis criterion of **presence or absence of conceptualization or definition,** what was found was that the authors conceptualize the content, making the information and explanations about the subjects explored clear, enabling a general understanding. Analyzing unit 4, which deals with water and health, it was observed that the authors present how diseases are transmitted, what the symptoms are and how to prevent them, but they do not define diseases, in other words, they only conceptualize them. In order to understand this comment, it is necessary to present the difference between defining and conceptualizing. According to Morh (1994),

> Concept' is used, [...], in its broader meaning of a general idea or notion about something. Thus, when it comes to conceptualization in a textbook, it is understood that the text should present information and explanations developed in such a way as to allow the student to understand or have a general conception (albeit imprecise and not formalized) about the subject in question. The 'definition', on the other hand, is more formal, more rigid. In order to define, you have to be precise: you have to delimit the object of the definition to its most individual and restricted form. For the conceptualization, the presence of the definition is not a *sine qua non;* this is generally false in the reverse situation: without the conceptualization, it is not always possible to understand a definition (p. 55).

This is a practice observed in both the 2nd and 3rd grade textbooks: the authors do not define the subject, but give general information about it.

Other times, the information is very superficial, insufficient or even unacceptable from the point of view of scientific correctness. This problem was also noted by Morh (1994) in his research and ends up representing one of the main problems and shortcomings of the Open Door Collection. To illustrate, let's look at a text from the book:

> **Neither plant nor animal kingdom**
> There are beings that don't belong to either of these two kingdoms. This is the case with fungi, bacteria and protozoa, which belong to different kingdoms.
> In addition to these, there are also viruses, which are not classified in any of the five existing kingdoms (GIL; FANIZZI, 2011, p. 110).

In this text, the authors leave the students confused. What are the five kingdoms? The book talks about the animal and plant kingdoms and includes the text above and a picture of a protozoan, a bacterium, a virus and a mushroom. This leaves a series of unanswered questions: Which kingdom do fungi, bacteria and protozoa belong to? What about viruses? Are viruses in the same classification as fungi, bacteria and protozoa? The information in the text is insufficient.

With regard to the **economic and geographical realities** criterion, the book addresses and reports on some problems that are occurring in different Brazilian regions - garbage in a stream in Sao Paulo (p. 58), a polluted river in Campina Grande (p. 64), a water treatment plant in Curitiba (p. 90), a garbage dump in Brasilia (p. 99), and even outside our country (oil spill in Korea). 64), a water treatment plant in Curitiba (p. 90), a garbage dump in Brasilia (p. 99), and even outside our country (an oil spill in South Korea (p. 75)), all of which meet the criteria. In this book, there are no illustrations or parts of texts that depict the state of Goias, so it will be up to the teachers to promote and stimulate reflection.

When analyzing the criterion of **suitability** for **the minimum age**, it can be seen that the content is suitable for children to understand. To illustrate this, below is a text from this unit on the rights and duties of citizens,

> You have learned that we have rights and that we can demand policies and actions from the public authorities, such as:
> - Drinking water supply.
> - Proper water and sewage treatment.
> - Collection and proper disposal of waste and sewage.
> - Selective waste collection and recycling.
> - We can't leave it up to the government to take care of the environment. It's up to us to contribute with individual actions such as:
> - Don't throw garbage out of the windows of cars, buses and trains into streets, streams or abandoned lots.
> - Separate glass, metal and plastic packaging and clean paper to reuse or send for recycling.
> - Pack garbage properly so that it doesn't spill.
> - Recover broken objects and toys and donate them.
> - Don't waste food, make the most of leftovers.
> - Avoid buying products that use a lot of packaging.
> - Use cloth bags when shopping.
> - Avoid disposable products such as cups, plates and forks.
> - Avoid wasting school supplies.
> - Making use of school supplies left over from the previous year.
> - Keeping the classroom and playground clean (GIL; FANIZZI, 2011, p.

99100).

In this quote, it can be seen that the information provided is easy to understand for children in the third year of elementary school. It emphasizes the government's responsibility for basic sanitation, but at the same time shows the individual's obligations towards the environment.

By teaching that children should not only claim their rights, but also try to do their part, the authors encourage readers (students and teachers) to reflect on their own practices and, at the same time, show the importance of taking care of the environment in order to live a healthy life.

In terms of analyzing the **environment**, it can be seen that the way the content is presented does little to explore the influence of the environment on health and disease conditions. As Mohr (1994) notes in his research, instead of presenting health conditions "as a dynamic state, dependent on the interactions that man maintains with the biotic, physical and social environments and the relationships existing in his own body" (p. 61), the emphasis is on disease (forms of contagion, symptoms and prevention). The same happens in the 3rd grade book from the Open Door Collection.

Few texts explain the influence of the environment on the health conditions of living beings. One of the few texts that explores this situation was found in Unit 5 - Separating materials from a mixture, when it reports on an oil spill at sea and the consequences for living beings - animals and seaweed. Below is the text of the book:

**Oil and water**

Oil is a material that does not dissolve in water. It forms a surface layer that prevents oxygen, air and sunlight from penetrating the water, making it difficult for aquatic animals to breathe and for seaweed to produce food. In addition, it sticks to the bodies of many fish, seabirds and other animals and can lead to their death (GIL; FANIZZI, 2011, p. 75).

Reaffirming what has already been said, the topic of health/disease cannot be analyzed considering only human beings, but living beings in general, which is why the example refers to the impact that oil spills in the sea can have on animals and plants.

In the criterion related to the **depth of** the content, the authors try to explore new approaches and aspects of a subject already covered in another unit within the same book and also complement and deepen the content covered in the 2nd grade book, from the same Collection. For example, the topic of water was covered in units 4, 5 and 6 of the 3rd grade book, and at the same time it expands on the information given in units 6 and 7 of the 2nd grade book.

The book presents different **activities**, which is why the expression indefinite appears in table 1. Some activities appear with the subtitle YOUR TURN (p. 99, 100) - an opportunity in which the student will have to answer the questions on their own, in which, normally, "they simply have to identify, in the previous text, the passage relating to the question proposed to them and complete the answer with the author's phrase" (MOHR, 1994, p. 61); therefore, it does not meet the criterion analyzed, which is to awaken the ability to analyze, criticize and position oneself. There are fewer activities that probe individually experienced situations in everyday life. There are also activities with the subheading IN DOUBLE (p. 69) or IN GROUP - these are exercises with a higher degree of difficulty, i.e. the text doesn't contain the answer, giving the student a little more reasoning to respond and be able to reflect, debate with one or more other classmates. With the subtitles INVESTIGANDO E EXPERIMENTANDO (p. 92-94, 112-114) or MANOS A OBRA (p. 100) -, the student will do a practical activity, or experiment, in which they can develop various skills such as observing, comparing, analyzing, reporting, drawing, graphing, among others; with the subtitle RECORDING IDEAS (p. 101) -, they are taken up again. 101), the main ideas worked on are taken up again, and the content is reproduced, but with the understanding that, when the student goes back to the text to re-read what has been written to find the answer, they may also be learning.

In addition to these activities, the book offers sub-titles such as READ TO KNOW MORE (pp. 68, 86), READ MORE TO BE INFORMED (p. 75), READ TO BE UP TO DATE (p. 61), HAVE YOU READ?, BE AWARE (p. 75), ADVANCES IN SCIENCE which present data and information from different regions of Brazil and even from other countries, as well as new advances in science, which complement the content being covered.

The book also has a subtitle, which is completely connected to the research subject, which is TIPS FOR HEALTH (p. 67, 82, 87). In this section, the authors warn of some precautions to improve or maintain health.

In terms of **eating habits**, there is a lack of texts encouraging students to eat better. There is only a reference to a table of nutritional information on a milk carton (p. 79), in a unit that is not about health/disease but about how food is mixed, explaining the process of separating food. And in another part of the book, an exercise which, when referring to food, presents a shopping list and the children have to classify the origin of the products on the list, whether they are of plant or animal origin (p.117). It does not specify the importance of good nutrition for a healthy life, eating habits and health promotion. This is in line with the analysis that Ilha *et. all.(2013)* identified in their research, the information gap. According to the authors, the subject of health is little explored and in a superficial way,

> [...] the textbook does not contribute significantly to the formation of citizenship, as it does not encourage the development of students' autonomy, made possible by factors such as: discussions on public health policies; students' recognition of the relationship between habits and lifestyles with health promotion; as well as their role in the community, as an individual-author of their health and also responsible for the well-being of others (ILHA *et. all.*, 2013, p. 113).

With regard to the **active lifestyle** criterion, in Unit 9 - Knowing the Human Body, it was expected that this principle would be identified. In fact, the unit describes the parts of the human body, talks about the characteristics that differentiate one human being from another, about human appearance, presents excerpts from the declaration of human rights, refers to disabled people and the difficulties they face and experience in everyday life, and finally shows the physical and mental changes that occur throughout a human being's life. All of this information is considered important for self-knowledge and respect for differences, but the criterion requires the practice of physical activities to improve health and avoid illness, and there is nothing in this book that leads to this understanding.

In the area of **reducing health risks**, the texts only refer to the prevention of diseases, but in this area it is expected that the book will address issues related to the prevention of drugs, domestic accidents, among others.

To conclude the analysis of the 3rd grade book, the illustrations are colored drawings and photographs. The book notes that the representation is out of scale and the colors do not correspond to the real tones (p. 63) or show a scale of size in the figure, to avoid misinterpretation by the students.

### 3.3.3 Analysis of the 4th grade book

As in the books previously analyzed, the 4th grade book also has 9 units arranged as follows: Food (Balanced food, Nutrients, Food groups, The food pyramid, Processed food, Artisanal food, Why does food spoil? Food preservation); Water composition and properties (Water on planet Earth, Water composition, The physical states of water, Changes in the states of water); Soil care (Soil uses, The Earth's surface, Changes to our planet's surface, Soil fertility, Soil preparation); The characteristics of the Earth's atmosphere (The atmosphere, Air and its properties, A force of nature, Pressure and altitude, Air temperature, Air humidity, Influence of the weather on everyday life); Classification of vertebrate and invertebrate beings (Similarities and differences between animals, Vertebrate animals, Invertebrate animals); Vital plant functions (Plants and solar energy, Photosynthesis, Transpiration, Parasitic plants, Plant reproduction, The flower from the inside, Plants that can harm your health); Food relations of living beings (Food for living beings, Decomposers, Food chain, Food relations in the sea, Food web, The world of microorganisms);Waste treatment (Waste production, Waste collection and disposal, Reusing waste, Recycling waste, Decomposition time of certain materials, Improper waste disposal); Investigating the past (The origin of the Earth, Division into continents, The emergence of life, The reign of reptiles, Fossils, The emergence of human beings). Of the nine units, none has the expression health or disease in the title as found in the books analyzed above. Only one subtitle was found in the 4th grade book, entitled Plants that can harm health, in unit 6 - Vital plant fungi.

An analysis of the content covered in the units shows that the subject is approached in an unspecific way, but that importance is given to issues linked to the preservation of health/disease, as in the units: Food, Classification of vertebrates and invertebrates, Vital functions of plants, Food relations of living beings, Waste treatment.

The PCN de Saude Brasil (1997) emphasizes the need to associate hygiene issues with food in order to be healthy and avoid contracting diseases.

> The direct link between hygiene and food needs to be emphasized. Recognizing the possibility of water and food being contaminated by faeces, chemicals and agrotoxins, as well as identifying contaminated water, food and objects as a source of disease are all components of preparing students for healthy eating (BRASIL, 1997, p. 108).

Analyzing the 4th grade textbook, we see in unit 1 - Food, that the text is meeting the requirements of the PCN on Health. The unit describes the importance of a balanced and varied diet, containing proteins, carbohydrates, fats, vitamins and minerals; the relationship between water and the functioning of the body; food groups; the relationship between fungi and bacteria with food and health; malnutrition. In the HEALTH TIP (p. 18), the book encourages the consumption of food, but also explains that it should be eaten after washing hands and food, and that the food preparation environment should be clean. The authors encourage children to eat properly, avoiding eating too many foods that are not good for their health and leaving out others that are important for their physical well-being.

During the analysis of unit 5 - Classification of vertebrates and invertebrates, it was noted that, when explaining the classification of invertebrates, specifically insects, the authors talked about issues that are common to children's daily lives and that are directly linked to health: how to avoid lice and how to fight them, and about dengue fever, a disease transmitted by the *Aedes aegypti* mosquito.

In Unit 6 - Vital functions of plants, the texts deal with health, when they refer to plants that can damage health. In Unit 7 - Food relations of living beings, the world of microorganisms is discussed, including where they can be found, where they are used (vaccines, as food) and the diseases they can cause. Unit 8 - Waste treatment deals with landfills, garbage, waste recycling and the impact waste has on the environment and on health.

According to the criterion **presence or absence of conceptualization or definition**, there are often elaborate concepts that provide clarity of understanding and the illustrations help in understanding and retaining the content. However, there are still concepts that are superficial, insufficient or even unacceptable from the point of view of scientific correctness, as observed in book 3.

With regard to unfamiliar terms, the authors highlight them in blue and direct the reader to look up the meaning of the word in the glossary on page 153 of the 4th grade book. This concern with unknown terms appears in all the books analyzed.

In terms of **economic and geographical realities**, the book contains content that is of the utmost importance for learning science, but does not specifically address a single economic reality or region to explore the content. For example, in Unit 8 - Waste treatment, the images (p.117) shown before the text are of the landfill in Sorocaba (SP) and waste recycling in Rio de Janeiro, but as in previous books, there is no mention of the state of Goias.

With regard to the criterion of **suitability for the minimum age**, it can be seen that this book has increased the amount of information in the texts and the use of scientific terminology, but they are still in line with the student's cognitive level.

Morh (2000), in his research on elementary school science textbooks, criticizes:

> [...] the failure to consider the environment in the conditions of health and illness. Instead of presenting them as a dynamic state, dependent on the interactions that man maintains with the biotic, physical and social environments and the relationships existing in his own body, the authors prefer to emphasize disease solely as a contagious entity and characterize health as the absence of disease or accidents (MORH, 2000, p. 102).

What we found when analyzing the **environment** criterion in this 4th grade book is that the same thing happened as Morh (2000): the authors describe more about the conditions of the environment in general, not making any connection with the subject of health/disease. The question of health is rarely explored in relation to man's interaction with the environment. An example found in the book in which the authors establish the relationship between man and the environment and the consequences for the health of living beings was identified in unit 8 - Waste treatment, when they talk about the inappropriate disposal of waste and the environmental impact and dangers and risks to which people who work there are exposed.

In terms of the **depth of** the content, it can be seen that, in the 4th grade book, the authors have taken care to follow up on the content introduced in previous books, delving into some topics such as water, soil, living beings, among others. However, with regard to the topic of health/disease, this is not perceived. For example, in Unit 2 - Composition and properties of water, at no point did they take up or deepen concepts previously worked on, such as polluted and contaminated water, nor did they explore water-borne diseases. The same

This observation can be made in relation to unit 3 - Caring for the soil, in which there was no discussion of the use of agrotoxics and their impact on health. In Unit 4 - The characteristics of the earth's atmosphere, when the authors address the subject of air humidity, they could have explored what care should be taken with health when the humidity is low and the most common diseases.

As the level of difficulty of the content has increased, it has also been observed that the **activities proposed** and **the types of activities** require the student to reflect more on situations within a context, request personal responses from the students, suggest the construction of graphs using the data collected, propose connections between related topics, encourage student participation by demonstrating the knowledge acquired.

When analyzing the **eating habits** criterion, it can be seen that there is a specific unit for working on food content, which focuses on healthy eating, explains the difference between processed and natural foods, talks about processes for preserving food and the importance of drinking water to preserve health/disease. In this unit, the authors recommend eating a balanced and varied diet. Understanding that not all pupils in public schools are able to balance their diet, introducing fruit, vegetables, olive oil, cereals, meat etc., because social inequality is so great, and the basics of food are missing from the daily lives of many families, the book could have proposed activities or even texts that would have made it possible to reflect on these differences, and even advise people to make use of parts of food that are often discarded by the population, but which are rich in nutrients, such as broccoli leaves. It should be emphasized that there is no single or standard diet; each individual and/or family must establish their own diet, according to the local, cultural and socio-economic reality.

According to the PCN on Health

> It is essential for the school to work together with the family and other reference groups for the student, taking into account available resources and established cultural standards. The concept of a "correct" universal diet should be avoided, as it could discourage the construction of a desirable dietary pattern compatible with the local culture, based on nutrient-rich foods specific to each reality (BRASIL, 1997, p. 108).

Even so, children need to be taught how to eat healthy food so that, when the opportunity arises, they are aware of what they are choosing. Also in this unit on food, they have inserted a chart showing the breakdown of foods, the nutrients they contain and their function in the body to highlight their importance for health.

As the PCN de Saude Brasil (1997) states, "students are expected to be able to describe the basic nutritional needs of the human body, indicating the appropriate foods for a nutritious menu, using the resources and food culture of their region" (p. 117). Therefore, students need to be able to adapt their diet to what they can afford, and not rely on foods that are impossible to find in the region where they live. In his analysis of this problem, Morh (2000) states that "it would be more useful to propose texts and activities that encourage students to explore their daily food habits, their successes, desirable changes and shortcomings" (p. 101).

In the 4th grade book, there are no illustrations or phrases in any of the units that highlight an **active lifestyle and** encourage physical activity.

When analyzing the criterion of **reducing health risks,** in Unit 5 - Classification of vertebrates and invertebrates, under the subtitle HEALTH TIPS, the authors explain how to avoid lice (p. 81) and, in another tip, how to avoid dengue fever (p. 86). Elsewhere, in Unit 6 - Vital plant fungi, there is a section explaining which plants can damage your health (p. 100).

### 3.3.4 Analysis of the 5th grade book

In the Open Door series, the four books analyzed were divided into nine units. In the 5th grade book, the units are as follows: Which way to go? (The Earth, Representing planet Earth, Imaginary lines, Cardinal points, The incidence of sunlight on the Earth, Bussola, Magnetic forces, The poles of the magnet and Earth's magnetic field); Taking care of water (Water everywhere, Water: a natural resource indispensable to life, Water pollution, Misuse of water, Importance of riparian forests, Source areas, Deforestation, Water-saving tips, Treatment plants, Untreated water, Water day); Soil and food production (The importance of soil, Burning, Cutting down trees, Monoculture, Erosion, Sustainable agriculture, Rotating pastures, Organic products, Transgenic products); Polluted air: the Earth in danger (Changing air composition, Environmental problems, Acid rain, Destruction of the ozone layer, Worsening the greenhouse effect and global warming, Combating air pollution, *Biodiesel);* Living beings and their relationship with the environment (Where living beings live, *Habitat,* Environment, Living beings and non-living beings in the environment, The various ecosystems, Cities, Living beings and non-living beings, Abiotic factors, Biotic factors, When an animal becomes extinct); Our body: organization and functioning (Cell, basic unit of living beings, Tissues, Organs, Systems); Human body: Regulation, reproduction and health maintenance (Nervous system, Sense organs, Endocrine glands, Male and female genital system, Fertilization, Menstruation, Sexually transmitted diseases, Healthy living); Types of energy (Energy in transformation, Production of heat, Combustion, Conductors of heat, Transformation of energy in animals, Transformation of energy in plants); Electrical energy (What is electrical energy, Hydroelectric power station, Wind power station, Thermoelectric power station, Nuclear power station, Electricity in motion, What was it like in the past? Dangers of electricity.

Of the nine units, only one refers to health in its title, unit 7 - human body: regulation, reproduction and health maintenance. However, according to the content analysis, six of the units deal with health/disease issues.

Unit 2 - Taking care of water, emphasizes the importance of this essential resource for the survival of living beings. It contains information on how to avoid wasting water, shows the worldwide problem of water contamination, the illnesses caused by drinking contaminated water (p. 29), the lack of preservation of water sources (p. 33), describes water treatment plants (p. 36), the difficulty of finding water, and also the Universal Declaration of Water Rights (p. 39).

In Unit 3 - Soil and food production, the authors talk about food production, the ways in which it is grown, the importance of preserving the soil (p. 43) and of burning (p. 44-45). The reference to health/disease can be seen when they comment on: sustainable agriculture and the production of healthier food (p.51); the topic of organic products, when they mention that food grown with agrotoxins is bad for your health (p.54); the consumption of organic food, which is richer in nutrients (p. 55).

In Unit 4 - Polluted air: the Earth in danger, the book describes what causes air pollution, has a table with two columns, one on what the pollutants are and the other with the main sources of pollution (p. 61), talks about respiratory problems caused by air pollution (p. 63), discusses environmental problems that occur, such as acid rain and the destruction of the ozone layer and how the sun's rays can harm health (p. 65).), discusses environmental problems that occur, such as acid rain and the destruction of the ozone layer and how the sun's rays can damage health (p. 65), refers to biodiesel as a non-polluting fuel alternative, but at the same time also comments on the problems it brings, especially the increase in food prices (p. 73).

In Unit 5 - Living beings and their relationship with the environment, the book discusses how living beings live, questions about the climate and climatic differences, the importance of sunbathing to produce vitamin D (p. 89) and of preventing excess sun and using sunscreen to avoid diseases (p. 89).

In Unit 6 - Our body: organization and functioning, the 5th grade book presents a study of: the cells, organs and systems (digestive, respiratory, urinary, cardiovascular, skeletal, joint and muscular) that make up the body; the importance of properly functioning systems in order to consider the individual healthy (p.107). The book confines itself to describing how the systems work and makes no mention of health/disease issues.

In Unit 7 - The human body: regulation, reproduction and maintaining health, the authors follow on from the previous unit and address issues related to the health of the body. The unit presents how the process of fertilization takes place (p.132) and what menstruation is (p.133); it describes the causes, treatment and ways of preventing sexually transmitted diseases, with a focus on AIDS and HPV (p.135-138); it emphasizes how to live healthily (p.139).

In Unit 8 - Types of energy, the book discusses the various types of energy, how man, through physical activity, can be producing energy. In the unit's health tip, it describes that even sleeping produces energy (p.150).

' While analyzing the content of the 5th grade textbook, as well as the books analyzed above, according to the criterion **presence or absence of conceptualization or definition**, it can be seen that the authors present some concepts that provide superficial, insufficient or even unacceptable information from the point of view of scientific correctness, such as when the authors talk about pollutants (p. 61), "What does AIDS have to do with me?" (p.136) and also about respiratory problems.

As in the previous books analyzed, this one also shows unknown terms highlighted in blue, and directs the reader to look up the meaning of the word in the glossary, which can be found on page 172 of the 5th grade book. Some words are highlighted in green and the meaning is given in a corner of the page. This concern with unfamiliar terms appears in all the books analyzed.

Looking at the **economic and geographic realities** criterion, it can be seen that the content of the book addresses various types of problems in different regions of the country. In Unit 2 - Taking care of water, the illustrations show: the spraying of a plantation in Maringa PR; the pollution of the River Tiete in Sao Paulo (p. 27); the issue of misuse of water in the city of Caninde de Sao Francisco -SE (p. 30); and when addressing the issues of untreated water, there is an image of a pond in Pirenopolis-GO (p. 37). Of the four books analyzed. Only this image from the state of Goias was found.

Gil and Fanizzi (2011) point out in this unit's HEALTH TIP that "Untreated water needs to be filtered and boiled to eliminate probable microorganisms that can cause illness. The use of chlorine is also effective in treating water" (p. 38). This is an extremely important measure to avoid diseases caused by drinking contaminated water.

When analyzing the criterion of **suitability for the** student's **minimum age**, the explanatory texts increased in difficulty, but always paying attention and respecting the student's cognitive level, seeking more complexity in the content. For example, unit 6 - Our body: organization and functioning, describes the inside of the body, the parts of the cells, the digestive system and the process of transforming food during digestion (p.109). The subject is more complex to understand, which is why it needs to be covered after the introductory process, which takes place in the previous years. The authors continue and go into more detail about the systems in Unit 7 - The human body: regulating and maintaining health, when they suggest reflecting on "what you can do to improve the health of your body" (p.122). Mohr (2000), in his analysis, criticizes the way in which the theme of health/disease is described in the collections, in a fragmented way, as if they were not interconnected.

> The splitting up of complementary and interdependent subjects and the mere presentation of facts would not be so serious if there was a proposal to integrate and summarize the content, before or after the individualized analyses [...]. The subject of *health,* for example, is divided into care for food, hygiene, basic sanitation and health problems. These subjects are treated in a watertight manner [...], as if the mutual relationships between them did not exist (MOHR, 2000, p. 94, emphasis added).

Understanding that the subject of the human body and health/disease are directly linked, this subject doesn't need to be separated, but can only go deeper into the subject, unifying the contexts.

The **environmental** issues covered in this book are man's interference in the environment, deforestation, air and water pollution, the destruction of the ozone layer, the problems they cause and how they affect man's health/disease process. An example that can be cited is that covered in unit 6, which talks about air and air pollution. In a proposed activity, the following situation is presented for the reader to reflect on: "Felipe lives in an industrial city, in Greater Sao Paulo, and almost always has a cough and burning eyes. Why do I always feel like this?" (p. 63). In this question, when the student answers what causes these symptoms in Felipe, he will be reflecting on the changes that the environment undergoes, whether immediately or in the long term, due to the action of man and the consequences and impacts on the health and quality of life of living beings.

In terms of the **depth of** the content, the 5th grade book follows on from the issues covered in previous volumes, and this is made clear by the fact that it takes up topics such as: water, air, soil, the human body, among others. As the grade changes, the authors are concerned with delving deeper into the content.

In the specific case of the content related to disease prevention, it can be seen that in the 2nd grade book, the diseases presented were chickenpox, influenza and mumps (p. 31), common diseases caused by viruses. In the 3rd grade book, the diseases explored were leptospirosis, which is caused by a bacteria (p.65), cholera and dengue, diseases transmitted by the Aedes *aegypti* mosquito (p.66-67). The 4th grade book tells us that plant pollen can cause respiratory diseases (p.103) and about plants that are harmful to health (p. 102). In the 5th grade book, sexually transmitted diseases are presented (p. 135). There is no repetition of content, always broadening the range of information, respecting the cognitive level of the students.

The **activities proposed** and the type of **activities** in this book suggest that teachers revisit content that was covered in previous units (p. 4446), recommend that students do research in books, magazines or websites on the subjects covered in the unit (p.49), provide questions that lead students to express their ideas, activities that ask for personal answers, suggest experiments and explain what happened.

Below are some proposed activities exploring the health/disease theme from the 5th grade book. In Unit 2 - Taking care of water, the activity YOUR TURN shows a photo of a river with dead fish and asks: "What consequences can a polluted river have for the population?" (p. 29). Below, the types of illnesses caused by drinking contaminated water are described.

In Unit 4 - Polluted air: the Earth in danger, the activity proposal INVESTIGATING AND EXPERIMENTING suggests that students carry out an experiment to see what air pollution looks like (p. 62).

In Unit 7 - The human body: regulating and maintaining health, one of the proposed activities requires students to have general knowledge, exploiting life experiences acquired in their daily lives.

Write in your notebook what the virus that causes AIDS is called.
Reread the text and summarize, in your notebook, the part that tells you how the virus can enter the human body, adapting the text if necessary.
Why is it important to use a condom during sex? Answer in your notebook.
Write in your notebook what can happen to people who use syringes and instruments from dentists, doctors and manicurists without sterilization.
Why can't you run the risk of being infected by an HIV-positive person when you play with them or use the same bathroom they use? (GIL; FANIZZI, 2011, p. 150).

Other forms of activity are those of personal response to the student's own acquired knowledge. "What can be done at school to ensure healthy coexistence?" (p.139). In the activity under the subheading MAOS A OBRA, it is suggested that students conduct a survey with an adult with the following question:

1. Person's name
2. Age

3. Sex
4. Profession
5. Do you consider your work a form of sedentary lifestyle? Why?
6. Do you do any physical activity? What do you do?
7. What led you to take up physical activity?
8. Do you consider it important to be physically active? Why? (GIL; FANIZZI, 2011, p. 140).

This activity will allow the student to interact about certain situations with other people in the family's daily life.

Looking at the **eating habits** criterion, during the analysis of the book few texts and images were found that explore this issue, most of which are found under the HEALTH TIPS subtitle, as can be seen in unit 6 - Our body: organization and functioning, when the authors say:

To get the most out of the food you eat, you need to eat slowly, chew well and be in a quiet environment.
While digestion is taking place, you shouldn't do activities that require physical effort, such as carrying weights, swimming, playing soccer, running or gymnastics (GIL; FANIZZI, 2011, p. 111).

And in unit 8 - Types of energy, when the authors describe the importance of rest as a source of energy:

Even when we're asleep, we use energy because our bodies don't stop working. However, during sleep, energy consumption is much lower. Sleep promotes rest for

> the body, which is why it is essential to get a good night's sleep, with approximately 8 hours of sleep (GIL; FANIZZI, 2011, p. 150).

Although the 5th grade textbook analyzed explores and highlights these important contents, it should be noted that, just as Ilha *et. all* (2013) found in their research, they also found that there is little information "about actions aimed at recognizing the student's role as an agent of their own health, [...] that motivate the exploration of the students' daily diet, the successes, the necessary changes and the shortcomings observed" (p. 114). Also according to the authors, science textbooks are restricted to informing

> [...] what functions are exerted by food, its constitution, its origins and what diseases are related to its deficiencies, with no texts being found that work on cultural habits, food preferences and socio-economic conditions (ILHA *et.all.,* 2013, p. 114).

Although the texts in the 5th grade book are very clear on the subject, they should have given more emphasis to the importance of good eating habits and preserving health.

With regard to the **active lifestyle** criterion, it can be seen that the book refers to practicing physical activities in order to achieve good health, as can be seen in the following example:

> If you have the opportunity to walk, cycle, play ball, run and play, take advantage of it. Exercise at least twice a week for an hour.
> By exercising, we release tension and relax our bodies and minds, which brings us well-being. When this happens, we are contributing to our body's physical and mental health.
> Always remember that everything should be done without exaggeration. Excesses can be harmful to your health, as your body is only prepared to withstand a certain amount of physical activity (GIL; FANIZZI, 2011, p. 140).

This HEALTH TIP mentions physical activity. However, in unit 7 - Human body: regulation, reproduction and health maintenance, the main focus is on the functioning of the body, starting with the study of the nervous system, the cranium, the spinal cord, the sense organs, the male and female genital systems and others, with the question of physical activity only being mentioned in the health tip.

When analyzing the criterion of **reducing risks to health**, it can be seen that some texts, such as the one in Unit 7 - The human body, mention how diseases are transmitted and prevention is left out, i.e. it is dealt with in a very superficial way, such as the sexually transmitted diseases discussed in the book.

> Sexually Transmitted Diseases (STDs) are diseases caused by various types of agents. They are mainly transmitted by sexual contact without the use of a condom with an infected person and usually manifest themselves as sores, discharge, blisters or warts (GIL; FANIZZI, 2011, p.135).

Next, the authors talk about AIDS and the Human Papillomavirus (HPV).

By pointing out these diseases, they describe how they are spread, the symptoms, the methods of prevention and how to treat them, focusing on encouraging prevention.

The topic of health is explored in various everyday situations in the "Open Door" collection of science textbooks, but separated into units dealing with hygiene, food and the human body. In this context, we can say that it follows the same way of presenting the content of the science textbook collections analyzed by Morh (2000), who came to the conclusion that "there is no synthesis text or activity for the student to understand health as resulting from the actions of nutrition, hygiene and the absence of diseases or accidents" (p. 93). She comments that this way of presenting health/disease in separate units makes it difficult to connect the issues with the actions that children practice in their daily lives.

After analyzing the 2nd to 5th grade science textbooks in the Porta Aberta Collection, in order to identify the contents explored and how they are dealt with, we tried to identify the concept of health/disease that the authors address in the books. We found that at no time do the authors refer to a magical religious view of health, i.e. the cure of diseases is not related or linked to belief in any religion. On the contrary, the texts talk about forms of contagion, prevention and treatment.

They explain some diseases and are concerned with highlighting the etiologic agent: viruses, bacteria, fungi, insects, among others. This excluded the concept of health/disease based on the holistic model, which believed that the cause of the imbalance (disease) was strictly related to the physical environment (stars, climate and insects).

Although the authors talk about diseases caused by water pollution, air pollution and environmental factors that can harm health, partly following the empirical-rational (Hippocratic) model, they do not restrict themselves to environmental factors, but also consider endogenous and social aspects; therefore, they are not based on this model.

By talking about the creation of the microscope and its importance in aiding scientific research and understanding the causes of diseases, the book shows the concern that the biomedical model presents, which is to identify, through scientific research, the etiological factors of diseases and, consequently, to propose curative and preventive measures. The texts don't go into the details of the scientific research processes, but, as previously mentioned, they do present the etiologic agent of the disease, the forms of contagion, prevention and treatment. They don't just focus on the etiological factor.

By portraying facts such as the lack of basic sanitation (water treatment, sewage, selective waste collection), adequate housing and hygiene conditions, which affect society's lifestyle and end up causing illnesses, the Collection follows the systematic model, which says that the causes of illnesses are no longer just natural but also social.

Summarizing the ideas, assuming that the texts do not link the cure of diseases to belief in any religion, that in order to explain health/disease they try to identify the etiological agent, justify the cause of the imbalance (disease) by considering endogenous and exogenous factors (environmental, social, cultural) and propose curative and preventive measures, it can be said that the authors follow the natural historical model of diseases (HND).

The HND model aims to monitor the entire health-disease process, seeking to understand the interrelationship between the agent causing the disease, the host, the environment and the process of disease development, including the different methods of disease prevention and control.

Considering the examples given above, it can be considered that the texts in the Collection, in a simple and very concise way, according to the student's cognitive level, talk about the causative agents of some diseases, the symptoms, forms of prevention, taking into account biological, social, cultural and environmental factors, in a procedural view, following the HND model.

## 4 FINAL CONSIDERATIONS

The health/disease theme is fundamental for children to learn about the types of disease that exist, how they become infected and what to do to avoid becoming infected. Since the textbook is one of the teaching resources used in the classroom and the topic of health is of paramount importance, it is hoped that the books will explore the theme.

Although different teaching resources coexist and can be used in the school environment, it is clear that the textbook has been, and still is, the main teaching resource in the classroom in many schools, such as those in Jatal. This is why this research is justified, with the aim of analyzing how health/disease is approached in the collection of science textbooks for the initial grades of elementary school, adopted in 2014, in the municipal network of Jatal-GO.

During the bibliographical research carried out, different conceptions of health were identified throughout human history - the magical-religious vision, the holistic model, the rational empirical model (Hippocratic), the Western scientific medicine model (biomedical), the systematic model and the natural historical model of diseases (HND) (procedural model). Given these different conceptions, the aim was to identify the concept of health conveyed in the books during the analysis of the "Open Door" Collection,

After studying the books from the 2nd to the 5th year of the Collection, based on the categories of analysis surveyed - scientific correctness, economic and geographical realities, adequacy to the minimum age, environment, depth, proposed activities, eating habits, active lifestyles and reduction of health risks - it was observed that the authors address many issues related to health and that the model of the Historical Natural Conception of Diseases (HND) prevails.

Throughout the analysis of the topic of health/disease, it can be seen, in comparison with other research carried out on the same topic - focusing on another group of science textbooks - that some of the problems identified previously are still present, such as the fragmentation of the content, the lack of depth in certain content and concepts that provide superficial, insufficient or even unacceptable information from the point of view of scientific correctness.

When analyzing the "Open Door" Collection, it is clear that some of the contents and approaches proposed by the PCN on Health have not been addressed, such as the participation of the State as a health promoter. The colleague does not clarify the responsibility of the government in building health centers and hospitals, distributing medicines, providing quality training for health professionals, improving people's quality of life and providing access to basic sanitation. There is also a lack of information on how to request services and who is responsible for providing them.

The analysis showed that all the books in the Collection contain conceptualizations rather than definitions of the health/disease theme. Again, a conceptualization is a text that provides information and explanations that enable students to gain a general understanding of the subject in question, even if it is not very precise or formalized. In the textbooks analyzed, the content was superficial, lacking the depth required by the definition, which can hinder learning.

With regard to the category of geographical reality, it was noted that the authors are concerned to present images and cite examples of problems and situations in different regions of Brazil and even in other countries; however, the state of Goias appears in only one book. However, it is worth noting that the problems presented can be discussed and compared with those experienced in Goias.

When it comes to adapting to the minimum age of the student, it can be seen that the authors use simple, easy-to-understand language that makes it easier for students to learn. As the grade of the book increases (4th and 5th grade), there is a noticeable change in the language of the texts.

All the lectures relate health/disease to environmental factors, but in a superficial way. The subject most explored refers to human actions and their impact on the environment, interacting with the health/disease process of humans and animals. It is worth noting that there is little exploration of fundamental issues, such as ways of preserving the environment

In the category of deepening the content, the colleague is concerned with giving continuity to themes explored in another unit of the same book and in the book of another grade. It is clear that some content is explored in greater depth throughout the lesson, such as water. Although this is the case, they could have explored more of the knowledge acquired in other grades with what is being presented at the moment, following up and going into greater depth on the issues presented.

The analysis of the activities in the colleague's book showed a diversity of proposals. In the 2nd and 3rd grade books, there was a certain proportionality between the activities that require memorization and the activities that awaken the student's reflection and critical capacity. In the 4th and 5th grade books, however, the activities demand more of the student's individual knowledge and, aware of the need to revisit previously studied content, they suggest that the answers in the activities should, in some cases, be personal, leading to reflection and the expression of general knowledge, making a connection between the content covered and the health/disease process.

During the analysis of the eating habits criterion, it was found that the books present the foods, but do not relate them to the importance of having hygiene habits and proper nutrition. Only the 3rd grade book does this, but in a superficial way.

In the 2nd, 3rd and 4th grade textbooks analyzed, reference was made to the issue of physical activity, but there was no connection to the importance of this practice for improving health. Only the 5th grade book suggests practicing regular physical activity for the body to function properly and achieve physical well-being.

All the books analyzed refer to the category of reducing health risks. Even if superficially, they try to relate the health/disease process to prevention methods to avoid contracting diseases. It's worth noting that although the colleague addresses some diseases, how they are spread and how to prevent them, some topics proposed by the PCN on health were not explored, such as preventing domestic accidents and addictions (cigarettes, alcohol, medicines, toxic substances).

Although textbooks seem to have the upper hand when it comes to the content they cover, they don't do so in a completely satisfactory way, and it's up to the teacher to make a careful analysis of the content presented and the pedagogical proposals put forward. Even knowing that the textbook has been approved by the

PNLD, it may have conceptual errors and it is impossible for it to contain all the information necessary for the teaching and learning of your students. It follows, then, that the teacher should not just stick to the content of the textbook; it is imperative that they seek out other sources, other materials, complementing the information contained in the books.

Working on the health/disease theme in the book has provided me with more information on the subject, making me realize things that go unnoticed in everyday life. After carrying out this research, I realized that, in order to achieve a healthy life, it is necessary to change some habits. Today, I try to control my diet and also practise physical activity, which I used to not do at all.

It is hoped that this study can contribute to future research and alert people working in the field of education to the importance of analyzing textbooks, taking care when selecting the colleague to be adopted in their school.

# BIBLIOGRAPHY

AMARAL, Ivan A.; MEGID NETO, Jorge. Science textbook quality: what defines it and who defines it? **Ciencia & Ensino**, Campinas, n. 2, p. 13-14, jun. 1997. Available at: <www.scielo.br/scielo>. Accessed on: September 17, 2014.

BACKES, Marli Terezinha Stein; *et all.* Concepts of health and disease throughout history: from an epidemiological and anthropological perspective. **Ver. enferm. UERJ**, Rio de Janeiro, p. 111-117, Jan/Mar. 2009. Available at: < www.facent.uerj.br/v17n1/v17n1a21.pdf >. Accessed on: May 19, 2014.

BALESTRIN, Maria Fatima; BARROS, Solange Aparecida Barbosa de Moraes. The relationship between the conception of the health-disease process and the identification/hierarchization of health needs. **Revista polidisciplinarica eletronica da faculdade Guaigara.** 2009. Available at: <www.portalfadesp.mentorhost.com.br>. Accessed on: June 15, 2014.

BASSO, Lucimara del Pozzo. **Study on the evaluation criteria of the 1996 and 2013 PNLD Science textbooks** . 2010. Dispomvelem : www.nutes.ufrj.br/abrapec/viienpec/resumosr0181-1artigos. Accessed on: October 3, 2014.

BRAZIL. Ministry of Education/Secretariat for Primary Education. **National Curriculum Parameters:** Health. Brasília: MEC/SEF, 1998. p. 249-260. Available at: <www.portal.mec.gov.br/seb/arquivos>. Accessed on: April 19, 2014.

BRAZIL. Ministry of Education/Secretariat for Primary Education. **National Curriculum Parameters:** Health. Brasilia: MEC/SEF, 1997. p. 101-127.

BRAZIL, Ministry of Education/Secretariat for Primary Education. **National Curriculum Parameters:** presentation of cross-cutting themes, ethics. Brasília: MEC/SEF, 1997. p. **25-30**. Available at: <www.portal.mec.gov.br/seb/arquivos>. Accessed on: May 21, 2014.

BUSQUETS, Maria Dolors; *et all.* **Transversal themes in education:** bases for an integral formation. 5 ed. Sao Paulo: Atica, 1999. 196 p.

CRUZ, Marli Marques. **Conception of health-disease and health care**. 2011. Available at: <www.fiocruz.br/biblioteca/dados>. Accessed on: April 19, 2014.

FREITAS, Elisangela Oliveira de; MARTINS, Isabel. **Conceptions of health in science textbooks**. Postgraduate program in science and health education. Center for educational technology for health. Federal University of Rio de Janeiro. 2008, p. 7-22. Available at: <www. portalfae.ufrj.com.br>. Accessed on: May 24, 2014.

GIL, Angela Bernardes de Andrade; FANIZZI, Sueli. **Open Door**. 1ª ed. Sao Paulo: FTD, 2011. 4 vol.

HOFFLING, Floisa M. Notes for discussion on the implementation of government programs: a focus on the National Textbook Program. In **Educagao e Sociedade**, Sao Paulo, v. 21, n. 70, p. 159-170, Apr. 2000. Available at: <www.scielo.br/scielo>. Accessed on: September 17, 2014.

ILHA, Phillip Vilanova *et all.* Health promotion in science textbooks from 6th to 9th grade. **Alexandria revista de educagao em Ciencia e Tecnologia** v. 6, n. 3, p. 107-120, novembro 2013 Available at: <www.alexandria.ppgect.ufsc.br/files/2013/11/Phillip.pdf>. Accessed on: May 14,

2014.

LEAO, Flavia de Barros Ferreira; MEGID NETO, Jorge. Official assessments of science textbooks. In: FRACALANZA, Hilario; MEGID NETO, Jorge (eds). **The Science textbook in Brazil.** Campinas: Komedi, 2006. p. 35-71

MARANHAO, Damaris Gomes. Children's health and well-being: a goal for early childhood educators in partnership with family members and health professionals. **Proceedings of the 1st National Seminar Curriculo em Movimento**. Belo Horizonte: Perspectivas Atuais, November 2010.

MEGID NETO, Jorge; FRACALANZA, Hilario. The science textbook: problems and solutions. **Ciencia e Educagao**, v. 9, n, 2, p, 147-157, 2003. Available at: <www.scielo.br/pdf/ciedu/v9n2/01.pdf>. Accessed on: September 17, 2014.

MOHR, Adriana. **Health at school**: analysis of textbooks for grades 1ª to 4ª . 1994. 99 p. Dissertation (Master's Degree) - Institute for Advanced Studies in Education, FGV, Rio de Janeiro, 1994. Available at: <www.scielo.br/pdf7ciedu/v6 n2 /02>. Accessed on: April 12, 2014.

MORH, Adriana. Analysis of the Content of "Health" in Textbooks. **Ciencia e Educagao**, v. 6, n. 2, p. 89-106, 2000. Available at: <www.scielo.br/pdf7ciedu/v6 n2 /02>. Accessed on: April 12, 2014.

PIMENTEL, Jorge R. Science textbooks: Physics and some problems. In: **Caderno Catarinense de Ensino de Fisica,** Florianopolis, v. 15, n. 3, p. 308-318, dec. 2006. Available at: <www.posgrad.fae.ufmg.br/posgrad/viienpec/pdfs/1608.pdf>. Accessed on: September 17, 2014.

NATIONAL TEXTBOOK PLAN (PNLD). Science textbook guides, 2010. Brasilia 2009. Available at: <www.fnde.gov.br/arquivos/category/125-guias?Dowloadpnld.ciencia>. Accessed on: October 3, 2014.

NATIONAL TEXTBOOK PLAN (PNLD). Science textbook guides, 2013. Brasilia 2012. Available at: <www.fnde.gov.br/arquivos/category/125-guias?Dowloadpnld.ciencia>. Accessed on: October 3, 2014.

REIS, Marcia Santos Anjo. **Science textbooks**: the environment as an investigated theme. 2000. Dissertation (Masters in Education) - Faculty of Education, Federal University of Uberlandia.

SABROZA, Paulo Chagastelles. **Conceptions of health and disease**. Sergio Arouca National School of Public Health. Rio de Janeiro, 2004. Available at: <www.abrasco.com.br>. Accessed on: May 19, 2014.

SCILIAR, Moacyr. History of the Concept of Health. **Physis:** Revista de Saude Coletiva, v. 17, n. 1, p. 1-4, 2007. Available at: <http://doi.org/101590/50103>. Accessed on: 07 Apr. 2014.

Printed by Books on Demand GmbH, Norderstedt / Germany